I0469728

Juvenile Firesetter Intervention Handbook

Prepared for

THE UNITED STATES FIRE ADMINISTRATION
FEDERAL EMERGENCY MANAGEMENT AGENCY

By

Jessica Gaynor, Ph.D.

For

SocioTechnical Research Applications, Inc.

Under Contract EME-97-RP-0015 from the Federal Emergency Management Agency, U.S. Fire Administration.

Statement of Purpose

This Juvenile Firesetter Intervention Handbook is designed to teach communities how to develop an effective juvenile firesetter intervention program.

The six chapters of this Handbook can be viewed as the six building blocks essential to construct a successful program. The cornerstone of the blueprint is understanding the personality profiles of juvenile firesetters and their families. The next step is identifying at-risk youth and assessing the likelihood that they will become involved in future firesetting incidents. The identification of the three levels of firesetting risk--little, definite, and extreme--leads to specific types of intervention, including evaluation, education, referral and follow-up. These are the critical components of a juvenile firesetter program. To provide a complete complement of services to juvenile firesetters and their families, the juvenile firesetter program must be part of a community network. This network consists of a continuum of care designed to provide a range of intervention services, including prevention, immediate treatment, and graduated sanctions to juvenile firesetters and their families. Finally, there is a specific set of programmatic tasks that will ensure the delivery of swift and effective intervention to at-risk youth and their families.

Target audiences for this Handbook are diverse. Because the fire service operates the majority of juvenile firesetter programs, these chapters are designed specifically to meet its needs. Fire educators, fire prevention specialists, arson investigators, juvenile firesetter program managers, interviewers, educators, and team members all will find this information useful. In addition, because juvenile firesetter programs are part of a community network, many other professionals working in the human service arena will benefit from the Handbook. These professionals include law enforcement, mental health, school personnel, child protective services, social services, and juvenile justice.

A planned and coordinated effort on the part of the fire service and human service organizations is the best way to reduce juvenile involvement in firesetting and arson and to protect and preserve lives and property in our communities.

FEDERAL EMERGENCY MANAGEMENT AGENCY

UNITED STATES FIRE ADMINISTRATION

FOREWORD

The United States Fire Administration (USFA), of the Federal Emergency Management Agency has funded and guided the development of this Juvenile Firesetter Intervention Handbook. The history of activity leading to this particular version dates back to two prior efforts. The first effort resulted in the publication of the USFA's three-volume <u>Juvenile Firesetter Handbooks (Ages Seven and Under, Ages 8-13, and Ages 14-18)</u>, 1978-1988. The second effort resulted in the collaboration between the Office of Juvenile Justice Delinquency and Prevention (OJJDP) and the USFA to produce the <u>National Juvenile Firesetter/Arson Control and Prevention Program</u>, 1994. Please see the following page for a complete list of contributors to these publications.

In 1997, a decision was made by the USFA to integrate the core content from the several different publications into one Handbook. The main author of this work is Dr. Jessica Gaynor, clinical psychologist, writer, and management consultant to juvenile firesetter programs. The Handbook was prepared under contract by SocioTechnical Research Applications, Inc. A working group, consisting of Dr. Gaynor, Dr. Patricia Hamilton, Cheryl Poage, and Dr. Manuel Gallardo, developed the content of the Handbook; and a review panel consisting of Captain Steve Cox, Marion Doctor, Dr. Kenneth Fineman, Carol Gross, Captain Kevin Kennedy, Irene Pinsonneault, and Paul Schwartzman provided editorial comment.

One of the unique contributions of the Handbook is the presentation of the Juvenile Firesetter Child and Family Risk Surveys. These surveys represent a shortened version of the Comprehensive FireRisk Evaluation. Both instruments can be used to classify youth and their families according to three firesetting risk levels. Juvenile firesetter programs will have the option of using a short assessment procedure with the Risk Surveys or a longer evaluation procedure with the Comprehensive FireRisk Evaluation. The development of the Comprehensive FireRisk Evaluation represents the original work of Dr. Kenneth Fineman with revision by the Colorado Juvenile Firesetter Prevention Program Project, consisting of Marion Doctor, Joe B. Day, Larry Marshburn, Kenneth Rester, Jr., Cheryl Poage, Paul Cooke, Carmen Velasquez, Dr. Michael Moynihan, and Elise Flesher. The Risk Surveys were the result of validation research by Moynihan and Flesher. Information about the development of these instruments can be obtained from the Colorado Department of Public Safety.

As with the prior works, the intent of this Handbook is to promote the development of a juvenile firesetter program that meets the requirements, and is sustainable within the resources of the community.

Information on how to acquire copies of this Handbook and about USFA-funded training may be obtained from:

United States Fire Administration
16825 South Seton Avenue
Emmitsburg, MD 21727
(301) 447-1189
www.usfa.fema.gov

CONTRIBUTORS

JUVENILE FIRESETTER HANDBOOKS

(AGES 7 AND UNDER, AGES 8-13, AND AGES 14-18)

Michael Baizerman, Ph.D.

Paul Boccumini, Ph.D.

Clyd Bragdon, Jr.

Gary Briese

Charles S. Brudo, Ed.D.

Esther S. Brudo, Ed.D.

Captain Joe B. Day

Beth Emsoff, M.S.

Kenneth R. Fineman, Ph.D.

Jessica Gaynor, Ph.D.

John Haney

Carl Holmes

David J. Kolko, Ph.D.

Lee Lewis

Cathy Lohr

Hugh McCless

Lynne Michaelis

Pat Miezala, R.N.

Tom Minnich

Connie Morris

Bill Peterson

Ray Waters

NATIONAL JUVENILE FIRESETTER/ARSON CONTROL AND PREVENTION PROGRAM

American Institutes for Research

Clifford Karchmer, M.A.

Institute for Social Analysis

Jessica Gaynor, Ph.D.

Acknowledgements

This Juvenile Firesetter Intervention Handbook was funded under a contract to SocioTechnical Research Applications (STRA) from the Federal Emergency Management Agency (FEMA), U.S. Fire Administration (USFA), (Contract EME-97-RP-0015).

There are many people who contributed their expertise to this Handbook. Andrew Giglio, the U.S. Fire Administration project officer, provided the initiative and guidance to see this project through to completion. Dr. Manuel Gallardo, President of STRA, and Dr. Patricia Hamilton designed the management structure for the project. This structure consisted of a working group and a quality review panel. Members of the working group included myself, Dr. Gallardo, Dr. Hamilton, and Cheryl Poage. The working group was responsible for developing the direction and content of the Handbook. Working group members also reviewed and edited drafts of the Handbook. Members of the quality review panel included Captain Steve Cox, Marion Doctor, Dr. Kenneth Fineman, Carol Gross, Captain Kevin Kennedy, Irene Pinsonneault, and Paul Schwartzman. Each quality review panel member provided content analysis, critical review, and editorial input. Katy Glad of STRA provided the technical support. Shayna York designed and executed the layout and graphics for the Handbook. Jim Post offered valuable editorial assistance.

The centerpiece of this Handbook is the presentation of two assessment and evaluation tools for interviewing juvenile firesetters and their families. Dr. Kenneth Fineman is responsible for the design and development of the Comprehensive FireRisk Evaluation. Many local fire departments supported its implementation. Cheryl Poage and the Colorado Juvenile Fire Prevention Program, funded by the Federal Emergency Management Agency, U.S. Fire Administration (EMW-95-S-4780), are responsible for producing the Juvenile Firesetter Risk Surveys, an alternative and shortened version for the fire service.

It is important to acknowledge that this Handbook represents the past quarter century's work of the U.S. Fire Administration (USFA) and the Office of Juvenile Justice Delinquency and Prevention (OJJDP) in the field of juvenile firesetting and arson. As such, it assimilates all the previous written documentation published by these agencies, including the USFA's three-volume <u>Juvenile Firesetter Handbooks (Ages Seven and Under, Ages 8-13, and Ages 14-18)</u>, and the OJJDP's five-volume <u>National Juvenile Firesetter/Arson Control and Prevention Program.</u> All of the local, state, and national fire service professionals and organizations, mental health professionals, law enforcement, juvenile justice and community service agencies need to be recognized for their contributions to these works. It is precisely these people who are responsible for promoting an effective prevention and intervention effort designed to reduce juvenile involvement in firesetting and arson.

Jessica Gaynor, Ph.D.
January, 2002

TABLE OF CONTENTS

Chapter 1

Juvenile Firesetters and Their Families

This chapter provides a comprehensive description of juvenile firesetters and their families. It is important to understand the development of various types of fire behavior in children and adolescents. Juveniles can be described according to their risk for future involvement in unsupervised firestarting and intentional firesetting. Key psychological and social factors describe the profiles of at-risk youth.

Objectives

1.1. **To define the four terms of fire behavior: fire interest, firestarting, firesetting, and arson.**

1.2. **To describe the three risk levels--little, definite, and extreme--of involvement in future firestarting and intentional firesetting.**

1.3. **To identify at-risk youth and their families using the factors of individual traits, social circumstances, and firesetting scenario.**

Fire Behavior

Fire behavior follows a naturally occurring development sequence in children. At least three distinct levels mark the chronological development of fire behavior in children: fire interest, firestarting, and firesetting. These categories of fire behavior represent increasing levels of involvement with fire. Through proper parenting, effective school and fire service education programs, and social interaction within their community, most children come to understand each of these levels of involvement and learn age-appropriate, fire-safe behaviors.

However, because of the influence of psychological and social factors, a certain percentage of children become involved in fire risk behaviors. Specific factors including emotional disorders, family dysfunction, and chronic stress can lead to such behaviors as unsupervised firestarting, and repeated, intentional, and malicious firesetting.

Fire Interest

Most children experience fire interest between the ages of three to five. Interest can be expressed in a number of ways. For example, questions about fire may be asked. These questions often focus on the physical properties of fire, such as how hot a fire is or what makes a fire burn. These questions are similar to questions children have about the other physical elements in their environment. For example, children may ask why the sky is blue or what makes water wet.

Children also express their interest in fire through their play. They may wear fire hats, play with toy fire trucks, and cook food on their toy stoves. This type of play is healthy and provides children with ways to explore and understand fire as a productive and useful part of their lives. It also represents the first signal to parents that it is time to educate their children about fire.

Firestarting

Firestarting occurs when children experiment with ignition sources such as matches and lighters. Most young boys between the ages of three and nine experiment at least one time with firestarting materials. If these events take place in controlled, supervised settings, most children learn age-appropriate fire behaviors. For example, if an eight-year old child requests and receives permission to help light the candles on his birthday cake, he can learn the conditions under which it is safe to strike a match and light candles. Children can be given the appropriate responsibility for fire at specific ages. Children should not be given this responsibility prematurely. The ages when specific fire-safe behaviors develop in children are discussed in a research paper listed in suggested readings at the end of this chapter. It is important for parents and caretakers to give children opportunities to participate in supervised fire-related activities so that they can learn how to conduct themselves in a safe and competent manner.

Unfortunately, many firestarts do not take place in supervised settings. A majority of children will engage in at least one unsupervised firestart. Most of these unsupervised firestarts are single episodes primarily motivated by curiosity. In general, fires resulting from these incidents are accidental or unintentional. They are started with available ignition sources and there is no typical target for these fires. If these fires get out of control, children will make an attempt either to put the fire out or go for help. The probability is low that a single-episode, unsupervised firestart will result in a significant fire. However, if children continue to participate in more than one unsupervised firestart, the probability of starting a significant fire increases dramatically. Therefore, it is very important not only to discourage unsupervised firestarts, but also to provide a solid education in fire safety to children to prevent unsupervised experimentation.

Firesetting

By the age of 10, most children have learned many of the rules of fire safety and prevention. They are capable of engaging in age-appropriate firestarting behaviors such as helping to light the family barbecue or building a campfire. If parents, caretakers, school, and fire service programs provide continuous guidance in fire-related activities, children will achieve a sense of competency and mastery over a powerful yet controllable aspect of their physical environment.

In some children, what begins as fire interest and leads to unsupervised firestarts, can result in repeated firesetting. Children in the age range of 7-10 years who understand the rules of fire safety may continue to be involved in repeated firesetting to pursue their interest, without their parents or caretakers being aware of this activity. While this firesetting is intentional, it may not represent any underlying psychological or social problems. This type of firesetting can lead to more serious incidents if parents, caretakers, or others who can provide the necessary guidance and education to stop the dangerous behavior do not discover it.

Intentional firesetting can be motivated by psychological or social problems. This type of firesetting consists of a series of planned firestarts that take place over several weeks, months or even years. The severity of these fires varies. They can range from parents finding burn marks and spent matches behind the house to fires requiring fire department suppression. Ignition sources, such as matches and lighters, are searched for, acquired, and concealed until they are needed. Firesetting usually takes place in an area near home, where there is little chance for immediate detection. Often there is an attempt to gather flammable materials, such as old newspapers or rags, or there is the use of accelerants, such as paint or alcohol, to hasten the spread of the fire. These fires can be motivated by a number of different reasons including anger, revenge, attention seeking, malicious mischief, crime concealment, and intention to destroy or harm property and/or people. The target of the fire frequently is specific in that it holds important emotional meaning for the firesetter. Once the fire is started, the firesetter will rarely make an attempt to extinguish it. Instead, the firesetter may run away to a safe spot, often to watch the fire burn. In some cases, the firesetter may run away and return later to view the destruction. There are even some reports of firesetters calling in the alarm and being first at the fire scene, volunteering their help with suppression.

Children and adolescents involved in unsupervised firestarting and intentional firesetting are the focus of this Handbook. **Table 1.1** summarizes the characteristics of unsupervised firestarting and intentional firesetting. These behaviors represent significant problems for these juveniles, their families, and their community.

Fire behavior naturally emerges in most children around the age of three.

Table 1.1
Firestarting and Firesetting

Factor	Firestarting	Firesetting
History	Single episode	Repeated
Method	Unplanned	Planned
Motive	Curious	Conscious
Intent	Accidental	Purposeful
Ignition Source	Available	Collected
Materials	At-hand	Flammable
Target	Nonspecific	Specific
Behavior	Extinguish fire	Run away

Arson

Unsupervised firestarting and pathological firesetting can be classified as a crime of arson if there is damage caused by a fire and it is determined that the juvenile involved acted recklessly or intentionally. This determination varies from jurisdiction to jurisdiction and from state to state. State statutes outline the specific circumstances that classify firestarting and firesetting as arson. The major element comprising most criminal-legal definitions of arson is intent.

If the juvenile's reasons for firesetting reflect substantial emotional immaturity or indicate mental illness, then it will be difficult to establish criminal intent. If the firesetting represents a conscious act to destroy, harm, or conceal another crime, then the behavior can be classified as arson. Intent includes both the purpose and design that motivates as well as a description of the mental state leading up to, during, and immediately subsequent to the firestart. A juvenile must have intended to participate in the act of firesetting with a mental state that is "sound" and "sane." If this type of mental responsibility is demonstrated, then firesetting can be classified as arson.

Typically a cause and origin determination is conducted to arrive at a decision as to whether an unsupervised firestart or intentional firesetting incident will be investigated as an arson crime. Chapter 2 focuses on how this cause and origin determination can lead to the decision of whether to pursue a legal or voluntary course of action for juveniles involved in firesetting.

Youth At-Risk

It is important to understand what types of children and adolescents become involved in unsupervised firestarting and intentional firesetting. Juveniles and their families can be described according to three risk levels. These risk levels represent the likelihood that youth will become involved in future firesetting. The three levels of risk are little, definite, and extreme. Each level of risk represents a successively more severe form of firesetting behavior.

The three risk levels can be described by the psychological and social factors of individual traits, social circumstances, and firesetting scenario. Individual traits are those characteristics describing physical, cognitive, and emotional functioning. Social circumstances refer to the quality of the family, social, and school environment. The firesetting scenario describes the behaviors and event leading up to, during, and immediately following the firestart. How these factors impact on the lives of juveniles and their families determines whether they will be classified as little, definite, or extreme risk for involvement in future firesetting.

Little Risk

Curiosity and experimentation motivate at least 60%-70% of those juveniles involved in unsupervised firestarting. Most of these children are at little risk for becoming involved in future incidents if they receive the proper supervision and education intervention. It is important to be able to identify and evaluate these children at an early age.

Table 1.2 describes the individual traits, social circumstances and firesetting scenario of little risk children. In general, the majority of these children are young boys between the ages of three and seven who come from all types of social and economic backgrounds. Young girls also participate in unsupervised firestarts, but they do so less frequently than their male counterparts. These youth do not exhibit significant psychological problems, their family and peer relationships are intact and stable, and their school performance and behavior meets expectations. The firestarts of little risk children typically are unsupervised incidents.

Table 1.2 Little Risk	
Factor	**Profile**
Individual Traits	The majority are young boys coming from a variety of social and economic backgrounds. Girls are involved less frequently. Physical, cognitive, and emotional development is normal. There is no evidence of psychiatric disturbance.
Social Circumstances	There is a functional family providing support and guidance. Peer relationships are adequate. School performance and behavior are well within the normal range.
Firesetting Scenario	Firestarts are unplanned single-episodes motivated by curiosity or experimentation. Resulting fires may be accidental. Available matches or lighters are used and there is no specific material or target ignited with the intention to destroy or harm. Attempts are made to extinguish the fire or call for help. Feelings of guilt or remorse occur after the incident.

Definite Risk

Between 30%-40% of children and adolescents identified with firesetting histories fall into the definite risk category. This category identifies juveniles who are very likely to engage in future firesetting incidents. The earlier these youth are identified, evaluated, and provided appropriate interventions, the better their chances of avoiding involvement in future firesetting. There are two major classes of definite risk juveniles, troubled and delinquent.

Definite Risk-Troubled Juveniles

The individual traits, social circumstances, and firesetting scenario for troubled juveniles are summarized in **Table 1.3**. Troubled juveniles can be described as the cry-for-help type in that by starting fires they intend to bring attention to their psychological distress. In fact, in most cases, it is their emotional conflict that motivates their firesetting. The source of this emotional conflict can vary greatly, and can include such things as family turmoil, abuse, neglect, unresolved difficulties in school, and other recent or chronic stressful life events. For these juveniles, firesetting is the symptom of their psychological pain.

While fire safety and prevention education may be helpful to these youngsters, it will not address their primary psychological problems. Therefore, once these juveniles have been identified and evaluated, in every case they should be referred to the appropriate mental health agencies. In addition to mental health services, there may be a need for other types of intervention, such as social services or juvenile justice. Referral to these community service agencies, including mental health, is covered in Chapters 4 and 5. If these youth and their families receive the help they need in a timely fashion, the chances are reasonably good that their firesetting behavior will not recur.

Intentional firesetting can be motivated by psycholo-gical or social problems.

Definite Risk-Delinquent Juveniles

Delinquent juveniles differ from their troubled counterparts in that they exhibit a certain pattern of aggressive, deviant, and criminal behavior. These behaviors emerge at a young age, and occur with greater frequency and intensity as the juvenile matures. For example, what begins as stubborn and disobedient behavior as a preschooler can lead to lying and stealing as a young child and to firesetting, petty theft, and vandalism as a teenager. Serious emotional or family dysfunction also may contribute to this pattern of antisocial behavior. **Table 1.4** summarizes the characteristics of the delinquent juvenile firesetter.

The longer this delinquent behavior pattern continues, the harder it is to reverse. Therefore, early identification is critical, but not always possible. Often these juveniles do not come to the attention of those who can help them until after they have set their first major fire. Fire safety education may impact these juveniles, but it will not turn them around. Depending on their histories and when they come to the attention of the fire service, they can be referred to mental health, social service, and other community agencies. Or, if their firesetting is classified as an arson crime, they can be referred to the juvenile justice system. Options for intervention with these juveniles are detailed in Chapters 4 and 5. This type of juvenile presents one of the biggest and costliest challenges to their families and their communities.

Extreme Risk

Less than 1% of firesetting children and adolescents fit this classification. These juveniles suffer from significant mental dysfunction. There are several severe mental disorders in which firesetting is a clinical feature.

These include the psychotic disturbances of schizophrenia and affective disorders as well as organically impaired disturbances of mental retardation and fetal alcoholic syndrome. Readings that contain more detailed descriptions of these severe mental disorders are recommended at the end of this chapter.

These severely disturbed children and adolescents are beyond most fire safety and prevention programs currently available. In fact, many of these children are a significant danger to themselves or others, and cannot adequately take care of themselves. **Table 1.5** summarizes some of the characteristics of this extreme risk level. If these youngsters come to the attention of the fire department, local mental health agencies should be contacted immediately.

Table 1.3 Definite Risk Troubled		Table 1.4 Definite Risk Delinquent	
Factor	**Profile**	**Factor**	**Profile**
Individual Traits	The majority are boys coming from a variety of social and economic backgrounds, although girls are also involved. One or more of the following problems exist: a greater number of physical illnesses, histories of physical or sexual abuse, poor impulse control, and overwhelming feelings of anger. For adolescent boys there may be gender confusion, higher levels of sexual conflict, lack of emotional depth, and greater risk-taking behavior.	**Individual Traits**	Most are boys, many of whom live in low-income households. There is female involvement to a less frequent extent. These young boys are impulsive, stubborn, mischievous, and disobedient. Preteens are generally defiant and frequently involved in lying and stealing. Teens are angry and aggressive and usually are involved in other antisocial activities such as substance abuse, petty theft, and vandalism.
Social Circumstances	Many live in single-parent households, with an absent father. There is little adult supervision and inconsistent methods of discipline. One or more parents may carry a psychiatric diagnosis. There are difficulties establishing and maintaining friendships. Learning difficulties are common, and attention deficit disorder with or without hyperactivity may be diagnosed. School performance and behavior are below average.	**Social Circumstances**	Many live in single-parent households, with an absent father. There is no formal supervision or discipline. Physical abuse and other violent patterns of family interaction are common. One or more parents may carry a psychiatric diagnosis; the most frequent is alcoholism. There is a small, but influential peer group that supports participation in antisocial activities. School truancy is typical, when school is attended, performance is poor and behavior is argumentative and defiant.
Firesetting Scenario	Recent or chronic stressful events trigger emotional reactions that result in firesetting. The fire represents the release of displaced emotions, such as frustration or anger. The fire also has the reinforcing properties of effect and attention. No attempt is made to extinguish the fire. There is no consideration of the negative consequences of the potential destruction.	**Firesetting Scenario**	Supported by their peer group, repeated, intentional firesetting occurs. Feelings of excitement and defiance are reported. Firestarting often is accompanied by other antisocial activities such as drug or alcohol use, petty theft, or vandalism. No attempts are made to extinguish the fire. Feelings of guilt or remorse are rare. There is little fear of the consequences or punishment.

Table 1.5 **Extreme Risk**	
Factor	**Profile**
Individual Traits	Depending on the diagnosed mental disorder, there can be a number of problems including extreme mood swings, uncontrolled anger, bizarre thoughts and speech, poor judgment, an inability to care for themselves, and potential harm to themselves or others.
Social Circumstances	Family background will vary according to the diagnosed mental disturbance. Often it will be difficult for these youngsters to live at home because of their impairment. They may be hospitalized, or live in a residential treatment facility. Peer relationships usually are poor. School performance is severely impeded by mental dysfunction.
Firesetting Scenario	Fire as a fixation may be a part of their mental disorder, therefore the reinforcing properties of the fires cause frequent firestarts. Reinforcing properties can be the sensory aspects of the fire or sensual or sexual arousal. Firesetting also may be a part of a delusional thought process. There is no rational or purposeful aspect to the firesetting, and the willingness to harm is difficult to predict.

Summary Points

- Fire behavior naturally emerges in most children around the age of three.

- If guided by parents, caretakers, schools, and the fire service, most children learn how to master and control fire in their environment.

- Because of the impact of psychological and social factors, a certain percentage of children become involved in fire risk behaviors that include unsupervised firestarting and intentional firesetting.

- The three levels of fire risk for juveniles and their families are little, definite, and extreme.

- The three levels of fire risk have corresponding personality profiles.

Suggested Readings

Topic	Resource
Child Development	Piaget, J. The Origins of Intelligence in Children. New York: International University Press, 1952.
	Erikson, EH. Identity, Youth and Crisis. London: Farber and Farber, 1968.
Child Development And Fire	Grolnich, WS et al. Playing With Fire: A Developmental Assessment of Children's Fire Understanding and Experience. Journal of Clinical Child Psychology. 14, 2: 128-135, 1990.
Attention Deficit Disorder	Cunningham, CE. A Family Systems Approach to Attention Deficit Hyperactivity Disorder: A Handbook for Diagnosis and Treatment. New York: Guilford, 1990.
Mental Disorder And Firesetting	Gaynor, J. Firesetting. In Lewis, M (ed.): Child and Adolescent Psychiatry. A Comprehensive Textbook. Baltimore: Williams and Wilkins, 2001.
Juvenile Delinquency	Loeber, R. and Stouthamer-Loeber, M. Development of Juvenile Aggression and Violence. American Psychologist. 53: 242-259, 1998.

Chapter 2

Identfication to Assignment

The heart of a juvenile firesetter program lies in its ability to identify and assess at-risk youth and their families. Once juveniles and their families are identified, then a preliminary assessment is conducted to determine a course of action. Assignment to the juvenile firesetter program involves screening the juvenile and family and having them participate in a formal intake process.

Objectives

2.1. **To describe how at-risk youth and their families come to the attention of a juvenile firesetter program.**

2.2. **To define the action options for assessing juvenile firesetters.**

2.3. **To assign the juvenile and family to the juvenile firesetter program based on a systematic set of procedures.**

2.4. **To outline a five-step screening procedure for a juvenile firesetter program.**

2.5. **To outline the intake process for a juvenile firesetter program.**

Identification

Typically, there are two ways juveniles involved in fire incidents come to the attention of a juvenile firesetter program. First, there are those juveniles who are referred from a number of different sources outside of the fire service, including parents, caregivers, schools, and community agencies such as law enforcement, mental health, child protective services, and various youth groups. These juveniles will display a range of behaviors from unsupervised firestarting to delinquent firesetting. Once they identify themselves to a juvenile firesetter program, a voluntary evaluation can be initiated to determine the severity of the fire incident.

The second way juveniles are identified is from within the fire service, generally as the result of suppression or investigation efforts. Immediately after suppression, most fire departments conduct a cause and origin determination. The purpose of a cause and origin determination is to learn as much as possible about the area of origin and how the fire started. This information is gathered by talking with firefighters at the fire scene, reviewing physical evidence at the fire, and interviewing witnesses. A cause and origin determination can identify juvenile involvement in a fire. In addition, observations made during suppression and investigation procedures can lead to the identification of juvenile firesetters. There are certain clues that suggest when youth may be involved in a fire. **Table 2.1** outlines some of the common characteristics of juvenile firestarts.

Once the fire service identifies the juvenile, the next action depends on a number of different factors. These factors include the nature and severity of the fire, the violations of local or state laws, the amount of sufficient evidence resulting from the cause and origin determination, the local operating procedures of the fire service, and the age and firesetting history of the juvenile. A youth identified by a cause and origin investigation can be referred to a juvenile firesetter program or to a juvenile justice agency depending on the juvenile justice guidelines.

Table 2.1
Common Features of Juvenile Firestarts

Location	**Younger children**-fires nearby home Outdoors: the yard Indoors: the bedroom **Older juveniles**-fires close to home Outdoors: wildlands and vacant lots Indoors: schools or recreation facilities
Ignition Source	**Younger children**-lighters and matches **Older juveniles**-lighters and matches with the addition of accelerants
Physical Evidence	**Younger children** Outdoors: burned paper or leaves Indoors: burned toys or clothing **Older juveniles** Outdoors: burned leaves Indoors: vandalism
Time of Fire	**Younger children**-most fires occur between 1:00 p m.-7:00 p m., with peak time between 3:00 p.m.-7:00 p m. **Older juveniles**-most fires occur between 10:00 p m.-1:00 a.m.

Assessment

There are two different types of assessment actions that can take place for juvenile firesetters and their families. They are legal and voluntary. The decision as to which action will take place depends upon a number of different factors related to the presenting fire incident. These factors include the violations of local or state laws, the losses (property, dollar or human) incurred from the fire, the amount of sufficient evidence resulting from the cause and origin determination, the local operating procedures of the fire service, the age of legal culpability, and the firesetting history of the juvenile. On a case-by-case basis, a decision to select either a legal or voluntary course of action will be made

taking one or more of these factors into consideration. In addition, there are certain cases in which both a voluntary course of action can be taken while legal actions are pending. This situation will be described in the following section on voluntary action. If the decision is made to take a legal action, then there are certain steps that must be taken on behalf of the juveniles and their families.

Legal Action

There are specific procedures to follow to ensure the integrity of the decision to take legal action. All of the legal guidelines and requirements of the particular jurisdiction and state must be followed. Initiating a legal action for firesetting is a very serious matter. At the point the arrest is made, all defendant civil rights must be recognized and honored. In many jurisdictions, juveniles and their families must be informed of the decision to arrest, and Miranda must occur. **Appendix 2.1** contains examples of Juvenile Miranda Rights statements. If further interviews with the juvenile are to take place, there may be specific legal requirements, such as parents being present, or approved interview locations, that must be recognized. Because policies and procedures vary from jurisdiction to jurisdiction regarding juvenile code, it is important for each juvenile firesetter program to consult with their local district attorney regarding the protection of a juvenile's legal rights.

Table 2.2 outlines four legal action options - citation, diversion, probation, and detention - for juveniles once their arrest occurs. Local law enforcement and the district attorney hold the responsibility for how the case is conducted. These responsibilities and procedures can vary from jurisdiction to jurisdiction and from state to state. After arrest, any interventions administered to juveniles and their parents are mandated by law. This often ensures that the necessary type of help and rehabilitation will be available. Arrested juveniles also are not necessarily excluded from participating in the voluntary actions described in the following section.

**Table 2.2
Legal Actions**

Citation

Juveniles can be issued a citation to appear before the probation officer at juvenile court. The probation officer may release the juvenile, keep the juvenile in custody, or release the juvenile to the family under house arrest with a promise to appear in court at a later date. The probation officer then investigates the case, and may decide to carry it no further, to release the juvenile without charges, or to refer the matter to the district attorney to initiate formal proceedings.

Diversion

Even though a crime has taken place, a decision is made not to take legal action. Everyone involved must agree to diversion. Diverting the case ends the possibility of further legal action for this offense. In these cases, there is always referral to alternative interventions. Firesetting cases that fall into this category can be referred to a juvenile firesetter program.

Probation

If the court elects to take legal action, one option for sentencing is probation. If the juvenile receives probation, he is likely to be released to his parents or legal caretakers. The juvenile may be ordered by the court to participate in a number of activities, such as a juvenile firesetter program or a community service program designed to help pay restitution.

Detention

This action is taken if the juvenile is in immediate danger or could cause immediate harm to someone else. If detention is taken, generally there will be an appearance in court within 24 hours to justify the action.

If local law enforcement or the district attorney decide that there may be a benefit derived from these voluntary procedures, then they also can be pursued. The specific juvenile justice intervention strategies that occur subsequent to arrest are described in Chapter 5.

Voluntary Action

If legal action or arrest is ruled out, then the remaining option is voluntary action. Doing nothing, sweeping it under the rug, or a slap on the wrist, are not acceptable responses to the fire incident. Legal action is typically ruled out when local or state laws are not violated, the cause and origin investigation results in insufficient evidence, and/or the juvenile does not meet the age of legal culpability. Most of the identified juvenile fire incidents will result in voluntary action. A voluntary action is defined as a decision not to file a criminal case and to take specific steps to ensure fire incidents do not occur in the future. The primary objective of a juvenile firesetter program is to prevent further juvenile involvement in unsupervised firestarting and firesetting.

The decision to take a voluntary course of action relies heavily on the cooperation of the juveniles and their parents. Unsupervised firestarting and firesetting will continue to occur if juveniles and families do not have access to the appropriate and necessary interventions. Because juveniles involved in fire incidents have been identified, a juvenile firesetter program is in the best position to provide an assessment of the severity of the fire behavior and to recommend a prompt course of action.

There are also certain instances when a voluntary course of action, that is participation in a juvenile firesetter program, can occur with a coinciding legal action. For example, a district attorney can recommend that a youth and family be referred to a juvenile firesetter program while their court date is pending. Or, the family may elect to participate in a juvenile firesetter program prior to their court date. And finally, the court can mandate youth and family involvement in a juvenile firesetter program. Therefore, voluntary entry into a juvenile firesetter program can occur independently or coincidentally with a legal action. In cases where both voluntary and legal actions are options, the juvenile firesetter program and the local district attorney must work together to deliver swift and effective intervention.

Assignment

Once identification and assessment is complete, effective assignment to a juvenile firesetter program involves systematic screening procedures and a formal intake process. The primary goal of screening is to allow the juvenile and family swift entry into the program. The intake process should be designed so that there is a secure pathway from the identification of cases to their assignment to the juvenile firesetter program.

Screening

A juvenile firesetter program provides a frontline attack by screening identified youth and their families. Because these juveniles have been identified, the fire service has created a significant but short window of opportunity to provide screening for these at-risk youth.

As soon as possible after the presenting fire incident, a juvenile firesetter program should initiate screening procedures. There are a variety of approaches that juvenile firesetter programs can take in setting-up their screening procedures. The selected approach will depend on several factors, including the structure and operation of the program and the availability of personnel and resources. These factors will be discussed in detail in Chapter 6.

Programs can use a five-step screening procedure that will provide youth and their families with rapid entry into the program. **Table 2.3** summarizes these screening procedures. Ideally, within 48 hours of the fire incident, the juvenile firesetter program should make contact with the juvenile and their family. In most situations, contact is made by telephone. There are circumstances in which the initial contact is made at the fire scene. If a determination is made at the fire scene that the

youth is a candidate for the juvenile firesetter program, the basic demographic information should be collected (see **Appendix 2.2** for examples of initial contact forms), and a telephone or personal interview appointment should be set-up.

Table 2.3
Five-Step Screening Procedure

1. **Initial Contact**

2. **Explanation of Program**

3. **Request Parent Participation**

4. **Information Collection**

5. **Refer for Personal Interview**

If the initial point of contact is by telephone, an explanation of the juvenile firesetter program is presented and parents are asked to participate in the program. If parents agree to participate, the program can proceed with information collection. The type of information programs collect over the telephone will vary, depending on their screening format and intake procedures. Some programs will collect a minimal amount of demographic information by telephone and ask the parents to bring their child in for a personal interview. Other programs will assess the severity of the presenting fire behavior over the telephone by asking specific questions or using a screening instrument. If programs use a telephone-screening instrument with a parent or caregiver, the Juvenile Firesetter Family Risk Survey is recommended (**Appendix 3.1**). This instrument is presented and described in detail in Chapter 3. After specific questions are asked or screening instruments are used, programs then can ask parents to bring their children in for a personal interview. An effective screening procedure will allow swift entry into the juvenile firesetter program.

Intake

Whether juveniles are identified by parents and caregivers, community agencies, or as the result of a cause and origin investigation, the juvenile firesetter program must have in-place a formal intake process. **Table 2.4** describes the basic steps comprising the intake process of a juvenile firesetter program. There are several questions that must be addressed by a juvenile firesetter program when it sets-up its intake procedures. First, where is the juvenile's initial point of contact? Will the juvenile enter the program through the fire service, and if so, how - through fire investigation, fire prevention, or contact with on-duty line personnel or firefighters? Second, what is a reasonable response time once the juvenile is identified to the program? The sooner the intake contact is made after the fire incident, the better the chance for a successful intervention. Third, who, in the program, will be responsible to make the initial contact? Will there be more than one person available to initiate the contact? There is a range of options; some programs have one contact person assigned per day, while others have one contact person available on a half-time basis. Fourth, what records will be kept of the initial contact? Will they be written or automated? At the minimum, an initial contact form should be filed. **Appendix 2.2** contains some examples of these documents currently in use by juvenile firesetter programs. Finally, what methods will be employed to prioritize cases? How will the program determine which cases are more urgent and therefore need to be assessed more rapidly than others? To build an effective set of intervention procedures, each juvenile firesetter program must develop its own answers to these critical questions.

Table 2.4 Intake Procedures	
Procedure	**Description**
Points of Entry	This is where the juvenile makes initial contact with the program. Some options within the fire service include fire investigation, fire prevention, or the local fire station.
Reasonable Response Time	The best window of opportunity is immediately after the fire incident.
Contact Person(s)	Identify the intake personnel and their availability.
Record of Contact	There must be a written or automated record of contact established for all cases. A sample Incident Report and Juvenile Firesetter Contact Form are contained in **Appendix 2.2**.
Prioritization of Cases	Some cases require a more rapid response than others. Methods for responding to urgent cases should be established.

A juvenile firesetter program must establish a secure pathway from identification to assignment. That is, there must be a standard route that all juvenile firesetter cases take from identification to assignment. For example, in one program when a juvenile firesetter is identified, an initial contact sheet is prepared and the paperwork is routed immediately to an on-duty shift liaison that assigns a firefighter to assess the case. In addition, all contacts are logged into a daily activity book. All juvenile firesetter cases are handled in this manner. A standard pathway from initial contact to assignment, which includes a paper trail, will ensure that no cases will be lost or will fall through the cracks. **Table 2.5** summarizes the typical pathway from identification to assignment for a juvenile firesetter program.

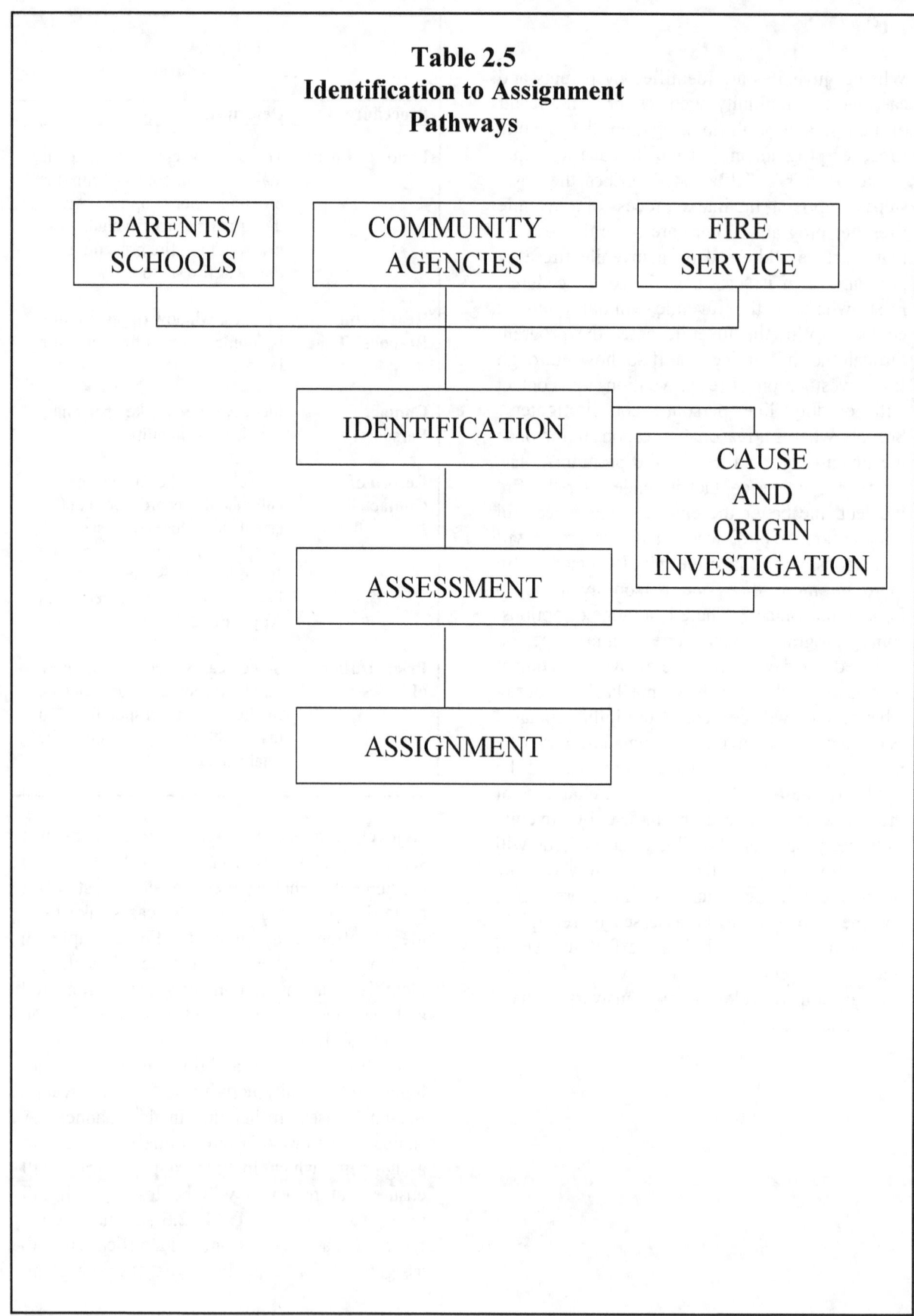

Table 2.5
Identification to Assignment
Pathways

Summary Points

- Parents, schools, community agencies, fire suppression, and fire investigation represent pathways used to identify juvenile involvement in firestarting and firesetting.

- Once juveniles have been identified as being involved in a fire incident, an assessment must be made as to whether to initiate a legal or voluntary course of action.

- A legal course of action for juvenile firesetters may involve arrest, citation, diversion, probation, or detention, depending on the age of legal culpability and the severity of the crime.

- Juvenile firesetter programs are primarily designed to work with those juveniles and their families who voluntarily seek help to prevent further involvement with fire.

- Assignment to a juvenile firesetter program involves a five-step screening procedure and a formal intake process.

Suggested Reading

Topic	Resource
A comprehensive listing of resources for community-based juvenile firesetter programs.	Juvenile Justice Clearing House. <u>The National Juvenile Firesetter/Arson Control and Prevention Program</u>. FA 145-149, 1994.

Chapter 3

Evaluation

The success of an intervention program for juvenile firesetters depends on its ability to accurately evaluate the firesetting problem. The method-of-choice for conducting an evaluation of firesetting behavior is the personal interview. Juveniles and their parents can be evaluated using a structured interview format. An evaluation interview yields a determination of firesetting risk as well as recommendations for intervention. If families participate in these procedures, the probability of involvement in future firesetting is greatly reduced.

Objectives

3.1. **To learn how to determine firesetting risk as a result of conducting evaluation interviews.**

3.2. **To outline the six essential elements of the evaluation interview.**

3.3. **To introduce the Juvenile Firesetter Child and Family Risk Surveys.**

3.4. **To present the Comprehensive Fire-Risk Evaluation.**

Risk Determination

The purpose of evaluating juvenile firesetters and their families is to determine their risk for involvement in future firesetting. The outcome of an evaluation is the classification of juveniles and their families into three risk levels. The three risk levels are little, definite and extreme, each representing a successively greater likelihood of future involvement in firesetting. **Table 3.1** describes the three levels of risk. For each level of risk there is a corresponding intervention. Juvenile firesetter programs report that roughly 60%-70% of identified juveniles fall into the little risk category, 30%-40% in the definite risk category, and less than 1% are classified as extreme risk. These percentages vary depending upon the level of active fire safety and prevention programs operating in the local schools and in the community.

The method-of-choice for determining firesetting risk is the personal interview. Using a structured interview format, juveniles and their families can be classified according to risk level. Once firesetting risk is determined, then recommendations can be suggested for

Table 3.1
Risk Levels

Little Risk	There is a small chance of a future firestart. The primary motive is curiosity or experimentation. The fire is probably not deliberate. Once the fire started, an attempt was made to extinguish it and/or call for help. There are feelings of guilt and remorse associated with the firestart.

Education intervention is recommend-ed. |
| **Definite Risk** | There is a significant chance of future firesetting. Motives can range from emotional problems to delinquent behavior. The fire was most likely deliberate. Once the fire started, there was no attempt to extinguish it. There was no call for help. The juvenile may have watched the fire burn and/or watched the fire service extinguish the fire. Feelings of guilt or remorse are usually absent.

Immediate referral to designated community agencies with education intervention as appropriate. |
| **Extreme Risk** | Another fire will occur. There is virtually no control over behavior in general, and firesetting in particular. There is imminent danger to person and place.

Immediate referral to designated com-munity agencies. |

intervention. There are two different instruments designed for use in the structured interview. They are the Juvenile Firesetter Child and Family Risk Surveys (**Appendix 3.1**) and the Comprehensive FireRisk Evaluation

(Appendix 3.2). Each of the appendices contains the instruments and complete instructions for their application and scoring. Both the Risk Surveys and the Comprehensive FireRisk Evaluation classify juveniles and their families into little, definite, or extreme risk for firesetting.

The method-of-choice for determining firesetting risk is the personal interview.

Comparing Interview Instruments

The decision of whether to use the Risk Surveys or the Comprehensive FireRisk Evaluation, or both, rests with each juvenile firesetter program. **Table 3.2** compares the application of the Juvenile Firesetter Child and Family Risk Surveys and the Comprehensive FireRisk Evaluation. There are several similarities between the instruments. The Risk Surveys were developed directly from the Comprehensive FireRisk Evaluation. Both instruments have the same approaches to developing rapport between the interviewer and the juvenile and determining the level of a young child's understanding of the interview questions. All of the questions contained in the Risk Surveys come directly from the Comprehensive FireRisk Evaluation and are validated within a 95%-99% range of confidence. Both instruments result in the same classification of little, definite, and extreme firesetting risk.

There also are significant differences between the Risk Surveys and the Comprehensive FireRisk Evaluation. Because the Risk Surveys represent a shortened version of the Comprehensive FireRisk Evaluation, the required interview time is significantly less for the Risk Surveys. The content of the questions on the Risk Surveys focuses primarily on firesetting behavior and the content of the questions on the Comprehensive FireRisk

Evaluation represents a broad range of content areas, including firesetting behavior as well as related psychological and social factors. Although the question on the Risk Surveys were developed from an analysis of responses to the questions on the Comprehensive FireRisk Evaluation, and they represent a 95% to 99% range of statistical accuracy, there is a small chance of miss-classifying juveniles into less severe risk categories. This chance exists because the determination of risk is being made on responses to fewer questions on the Risk Surveys than on the Comprehensive FireRisk Evaluation. Therefore, there may be juveniles and their families classified as little risk, when, in fact, they are at definite risk for involvement in future firesetting. To adjust for this, these borderline cases, that is, those cases falling between the little and definite risk categories (as indicated by a specific range of scores on the Risk Surveys), should be recommended for educational as well as mental health intervention.

The Risk Surveys and the Comprehensive FireRisk Evaluation provide a structure for conducting the evaluation interview. Before discussing these instruments in more detail, it is important to understand how to conduct such an interview and what factors contribute to a successful exchange of information between the interviewer, youth, and family.

	Table 3.2 Comparing the Risk Surveys and the Comprehensive FireRisk Evaluation	
	Juvenile Firesetter Child and Family Risk Surveys	Comprehensive FireRisk Evaluation
Purpose	Screening/ Evaluation	Evaluation
Method	Structured interview	Structured interview
Questions	Firesetting behavior	Firesetting behavior Psychological behavior
Design	Child Survey Family Survey	Child evaluation Family evaluation Parent questionnaire
Action	Brief interviews with child and parent	Extended interview with child and parents Parents complete questionnaire
Time	30-45 minutes	60-90 minutes
Outcome	Firesetting risk level; recommendations for intervention; recommendations for referral	Firesetting risk level; description of psychological behaviors; recommendations for intervention, and referral

The Interview

A structured interview is the method-of-choice for collecting evaluation information on juvenile firesetters and their families. A structured interview consists of a series of questions and answers designed to gather information. Conducting a structured interview with juveniles and their families as soon as possible after

firesetting will yield not only the facts and circumstances of the incident, but also the attitudes, behaviors, and levels of understanding of those being evaluated. There are six essential elements that comprise the structured interview. They are the target populations, the interview format, the interview style, special situations, confidentiality, supplementary interviews, and legal issues.

The Target Populations

The three target populations--the young child, the preadolescent and the adolescent--are primarily distinguished by age. Each age category often requires different considerations when conducting a structured interview. The special circumstances that can arise when interviewing these three age groups will be discussed throughout this chapter.

The Interview Format

The interview format consists of those factors that influence how the interview is conducted. These factors include the location, setting, scheduled time, and sequence of the interview. Each of these factors must be considered and planned prior to the start of conducting the interview.

Location

The basic rule is to conduct the interview in a place where there is a balance between comfort and support and the ability to maximize information exchange. Not only must the interviewer feel confident, but the juvenile and family also must feel secure enough to share information. Each juvenile firesetter program must decide where to conduct their interviews.

There are several location options. Many juvenile firesetter programs conduct interviews at the neighborhood fire station. A second option, if available, is an office space located in the fire department, but not necessarily at a fire station. For example, in urban areas, many fire departments have separate administrative offices that do not house fire suppression equipment. In some instances, depending on the nature and extent of the fire incident, interviews are conducted at the fire scene. Some juvenile firesetter programs conduct interviews in the home. Finally, juvenile firesetter programs have the option of selecting more than one site for the interview. For example, part of the interview may be conducted at the fire station, and part of the interview may be conducted in the home. The section of one or more interview locations rests entirely with the juvenile firesetter program.

Setting

Once a location is selected, there are several features within the setting that should be taken into consideration. If the interview takes place in the fire station or offices of the fire department, then the type of room and the way is which the room is set-up must be planned.

If the juvenile and family come to the fire department, it is recommended that a specific room be set aside to conduct the interview. Ideally, this room should have a door that closes firmly. If possible, there should be an area outside the room that contains some chairs and a table. Inside the room, there should be a set of chairs arranged in a semi-circular pattern. This arrangement is suggested so that all participants in the interview can see one another without interference. If an interviewer sits behind a desk, the desk can act as a physical barrier to communication. The semi-circular pattern creates an open seating arrangement and facilitates open communication.

Structured interviews conducted at the fire scene depend on the type of site and the nature and extent of the damage. Sometimes observing the reaction of the juvenile and family to the fire scene can yield a great deal of information about the firesetting behavior. The choice of whether to conduct an interview at the fire scene depends on the circumstances of each case and whether there is anything at the site that can be used to gain further insight into the firesetting behavior of the juvenile.

Interviews conducted in the home can provide additional information about the juvenile and

family that might not be observed during an interview in an office setting. If the home is the only interview location, when arranging the interview with the family, it is important to clarify that a certain amount of uninterrupted time must be set aside. If there are too many disruptions in the interview, the quality of the information collected during a home interview may not be as good as information collected in a more controlled office setting. In addition, it is recommended that, if possible, when visiting the home, two people conduct the interview. A team approach lends a certain amount of comfort and security in the home setting.

Scheduled Time

A structured interview requires that a specific amount of time be scheduled between the interviewer, the juvenile, and the family. The average amount of time for the structured interview will vary according to the type of instrument selected to conduct the interview. The range of interview time can vary from 30 to 90 minutes, depending on how the interview proceeds.

The interviewer should call and make a scheduled appointment with the juvenile and the family. The appointment should specify where the interview is to take place, and the specific time and day. The day before the interview, a reminder call should be placed to the family.

Sequence

The interview sequence refers to the order in which interviews are conducted with juveniles and their parents. All family members come to the same scheduled interview, but there is a suggested sequence for interviewing the family. The recommended sequence is first to interview the entire family, then to speak individually with the parents, followed by a separate interview with the juvenile, and to conclude with a final summary that includes the whole family.

This sequence allows interviewers to begin by getting to know all of the family members and to explain to them what they can expect from the interviews. In addition, it allows for some separate time with the parents and with the juvenile to get their individual perspectives on the presenting problem. In concluding the interview, it is important that the family gain a sense of closure and understanding about what has happened and how they are likely to be helped.

Recording the Interview

It is important to collect and maintain adequate documentation of interview information. Each juvenile firesetter program must decide how this will be accomplished. There are several options. Interviewers can take notes during the interview. However, communication is enhanced when an interview is conducted as a conversation, and note taking is kept to a minimum. Notes can be written immediately after the interview, when the scoring procedures are executed. The interview also can be audio or video taped. If this occurs, it is recommended that the juvenile firesetter program have parents sign a written consent form. Most often, parents do not object to recording the interview. Very little, if any, data is distorted. In fact, a more accurate record of the interview for future reference is contained on the tape. The question of how these records are stored and who has access to them is covered in a section to follow on confidentiality.

Interview Style

The interview style refers to the manner in which the interview is approached and presented to the youth and family. The purpose of a structured interview is to learn as much about the juvenile and family so that the interviewer can make an informed decision about how best to help the youth. This is not an interrogation-style interview where pressure is applied to receive answers to questions that can lead to a legal action. Rather, this is an informational interview where answers to questions can lead to help in resolving the problematic firesetting behavior. There are certain interviewer actions, attitudes and expectations, methods of establishing rapport and determining the level of understanding, procedures for asking questions, and ways of listening to responses that will help interviewers conduct a productive evaluation.

The Interviewer

One of the first questions a juvenile firesetter program faces is who will be conducting the interview. There are many factors to take into consideration in the selection of interviewers, including previous experience, ability to relate to juveniles and families, and teaching skills. A workshop or training program is recommended for all those who are selected as interviewers. This is discussed in more detail in Chapter 6.

Interviews can be conducted by one or two persons. The decision to use one or two interviewers rests primarily with the amount of resources available to the juvenile firesetter program. The major advantage of using two interviewers is that it creates a team approach, and can increase the amount of observation and information that is gathered during the interview. However, the one interviewer method has been used successfully by the majority of juvenile firesetter programs that conduct structured interviews for evaluation.

There are differing opinions about whether interviewers should wear uniforms or street clothes to conduct interviews. Some fire departments mandate the wearing of uniforms at all times. Some interviewers feel more comfortable wearing their uniform. Uniforms suggest that the interview is official and the person wearing it is recognized as the authority. Some interviewers suggest that wearing a uniform is more effective with younger juveniles while others suggest it generates fear in young children. Some interviewers think that a uniform creates a barrier when interviewing preteen or adolescent firesetters while others think is reinforces authority. Each juvenile firesetter program must develop their own policy regarding the dress of interviewers.

Expectations

Everyone participating in the interview comes with certain attitudes and expectations. Interviewers, juveniles, and parents will bring to the interview their hopes and ideas. **Table 3.3** presents some common but often unrealistic expectations that juveniles, parents, and interviewers bring to the interview. It is helpful if the interviewer is aware of these potential expectations, so that a more reasonable picture can be painted of the interview and its outcome.

Rapport

One of the interviewer's first steps is to create a safe and secure environment in which the youth and family are willing to share information during the interview. To accomplish this, the interviewer can think of himself/herself as an advocate, or acting on behalf of the juvenile and family. An adversarial, or antagonistic attitude will distance the youth and family.

```
┌─────────────────────────────────────┐
│          Table 3.3                    │
│   Common Interview Expectations       │
│                                       │
│ Source        Expectation             │
│                                       │
│ Juvenile      Confusion and uncertainty│
│               about what is going to happen│
│                                       │
│               Fear of punishment      │
│                                       │
│ Parent        Sense of relief to share the│
│               burden                   │
│                                       │
│               Someone else will solve the│
│               problem                  │
│                                       │
│               Punishment for the child │
│                                       │
│ Interviewer   Authority- "I'm in charge and I│
│               know what's best"        │
│                                       │
│               Substitute parent- "This is how│
│               to treat your child"     │
│                                       │
│               Rescuer- "I am here to solve all│
│               your problems"           │
└─────────────────────────────────────┘
```

What the interviewer does and says is critical for building communication with the juvenile and the family. The juvenile and family may be tense or anxious when talking about the fire incident. The interviewer can help them relax by talking about general, neutral topics such as the weather, sports, television programs, etc. The first section of both the Risk Surveys and the Comprehensive FireRisk Evaluation have a

series of questions titled, Development of Rapport. All or some of these questions can be used to help establish open communication channels at the start of the interview. The interviewer must deliver the message that they want to hear what the youth and family has to say and that they are there to listen to them.

Asking the Questions

The Risk Surveys and the Comprehensive Fire Risk Evaluation present the specific questions or content areas to be asked during the interview. These instruments are intended to serve as a structured guide for the interview. The objective is to be able to score the responses to the questions contained on these instruments at the end of the interview.

Although the instruments present a set of structured questions for the interviewer to ask the juvenile and family, many of the questions may need further explanation, may lead to other questions or to other topics of conversation. This can be a very positive feature of the interview, because more information is likely to be shared. It is important for the interviewer to remain open about exploring more questions, while at the same time keeping in mind the specific structure of the interview.

Attentive Listening

How the interviewer responds to the answers offered by the juvenile and family will set the tone of the interview. The interviewer should practice attentive listening. Attentive listening is showing interest in what is communicated by the juvenile and family. Interest can be shown by what is said as well as by what is done. **Table 3.4** outlines ways in which the interviewer can listen to the juvenile and family in a manner that will communicate interest and concern.

Special Situations

There are several special situations that can occur while interviewing juveniles and their families. Although the interviewer cannot be prepared for every unusual event, there are some situations that can be anticipated and handled in an effective manner.

Young Children

Interviewing children under seven is very different than working with older children. Because of their mental and emotional development, specific interview skills are necessary to conduct an effective evaluation.

Table 3.4
Attentive Listening Methods

Attentive Listening	Method
- Show Concern	Eye contact
- Listen Carefully	Do not interrupt, let them complete their thoughts
- Wait-Think-Response	Reflect before responding
- Repeat for Clarity	Recount in your own words what you hear
- Sharing of Self	Disclosing a small part to make a human connection
- Be Honest	Be truthful about circumstances
- Give Hope	Comfort, do not give false hope
- Observe Incongruities	Note differences between what is said and the body language
- Suspend Judgment	See the problem through eyes of the child and family

---

At the beginning of the interview, it is important to determine the level of understanding of a young child. Both the Risk Surveys and the Comprehensive FireRisk Evaluation have sections titled, Determine Level of Understanding, that suggests ways to work with a young child at the beginning of the interview. This section not only helps the interviewer determine the level of understanding, but using special interview tools, such as toys, games, and puppets, helps to increase the rapport between the interviewer and the child. It is important to complete this section of the evaluation for children under seven. This section will evaluate whether a young child understands what you are asking them. If they do not, the interview is not likely to be productive, and the interviewer may elect not to proceed. The interview with the parents, however, may be more helpful.

Language

A language barrier may exist between the interviewer, the juvenile, and the family. A determination should be made prior to the interview if English is not the first language of the juvenile and family. If not, an interpreter should be present to speak during the interview. If this is not possible, then a bi-lingual family member, if available, can help with the translation. In certain areas of the country where Spanish, Cantonese or other languages are common, efforts can be made to recruit interviewers who can speak these languages. In addition, if possible, it is useful to identify a sign language interpreter in the community who can help when interviews are conducted with hearing impaired children or parents.

There may be some interview circumstances where English is spoken, but the vocabulary and level of understanding may be limited. In these cases, the interviewer may need to adjust their own vocabulary and rephrase the interview questions.

A Spanish version of the Comprehensive FireRisk Evaluation is being developed. (The Spanish version of the Parent FireRisk Questionnaire is presented in **Appendix 3.1**). It will be distributed to those communities requesting it from the United States Fire Administration.

What the interviewer does and says is critical for building communication with the juvenile and the family.

Resistance

Resistance refers to a lack of cooperation on the part of the juvenile or family. Resistance can occur any time after contact is made between the juvenile firesetter intervention program, the youth, and the family. Certain situations can be readily identified as resistance, and there are steps that can be taken to work through these difficult situations.

Resistance on the part of the juvenile may be expressed in terms of not wanting to talk, lying, sarcasm, hostility, anger, or rudeness. Some children may even be playful and crack jokes, trying to divert attention from talking about the firesetting problem. This type of behavior can represent many things, including fear or anxiety about being interviewed, or it may be masking some underlying problem or conflict. Through questioning, the interviewer may be able to identify the concern. Once the problem is recognized, often the interview proceeds without further disruption.

If the resistance continues, and the interviewer is unable to talk to the juvenile, the interviewer has several options. One can return to rapport building, and spend more time developing lines of communication. Then the interviewer can make a second attempt to talk about the firesetting problem. The interviewer also can ask another interviewer to help out, reschedule the interview, or refer the juvenile to a mental health professional for further evaluation. This decision will vary from case to case and rests entirely with the interviewer.

Parents may express their resistance by making it difficult to set-up an interview, by skipping the first scheduled interview, or by not wanting to talk once they arrive at the interview. Parental resistance is difficult, and sometimes arouses anger in the interviewer. It is therefore very important for the interviewer to recognize this anger if it occurs and to understand its source.

If the parents are making it difficult to set-up the first interview, it is good to remind them about the serious nature and potential consequences of the firesetting behavior of their child. This usually is convincing. If parents miss the first scheduled interview, call them back and reschedule. If they fail to make the second interview, it is highly unlikely that they will give you their cooperation.

When parents are unwilling to talk during the interview, it is usually for a reason. Sometimes through questioning, the interviewer can pinpoint the problem. Parents may bring their own fears and concerns into the interview. Other times, the resistance represents a deeper set of conflicts that the interviewer cannot touch. If this continues to interfere with the interview, the interviewer has the same options as mentioned previously for the juvenile interview. It is important not to continue in an interview situation when it is clear that no information is going to be shared that will help the firesetting problem.

Abuse

In most states there are laws mandating those who suspect or identify physical or sexual abuse to report their observations immediately to the appropriate child welfare agency. Interviewers may become involved in these situations. It is important for each juvenile firesetter program to follow its state regulations and procedures regarding the recognition and reporting of physical and sexual abuse. **Table 3.5** describes some of the warning signs of physical and sexual abuse. There must be guidelines set-up for interviewers so that if they suspect or recognize abuse they will know exactly what they need to do and how they are to report it.

Severe Mental Disorder

In rare situations signs of severe mental disturbance may emerge during the interview. The three major types of severe mental disorders most likely to be observed are psychosis, depression, and suicide risk. Psychotic juveniles are not able to communicate their ideas or thoughts, they cannot answer simple questions in a straightforward manner, and they can appear either severely withdrawn or overly excitable. They may have distorted thoughts, delusions, or hallucinations. Depressed juveniles are tearful and cry without reason, they can be excessively irritable and angry toward their family and friends, and they often withdraw from participating in their usual activities. If a juvenile is at risk for suicide, they have repeated thoughts about death; and they may have a specific plan of where, when, and how they are going to hurt themselves. Identification of any one of these signs of severe mental disorder must result in the immediate referral of the juvenile and family to mental heath services.

When parents are unwilling to talk during the interview, it is usually for a reason.

Criminal Behavior

A situation can occur during an evaluation interview in which a juvenile or family member discloses new information about having committed a criminal act that may be arson, or another type of crime. A juvenile firesetter program should have a set of guidelines to follow if this situation arises. These guidelines will vary depending on both the legal code of the jurisdiction as well as who is conducting the interview. For example, if a youth claims to have set a school fire in the community, and if the interviewer is an arson investigator, the interviewer may elect to shift the focus of the interview from evaluation to custodial. In many jurisdictions, at this point in the interview the arson investigator will have to Mirandize. If the interviewer is not an arson investigator or a member of law enforcement, then the interviewer can continue to collect evaluation information and subsequently turn it over to arson investigation or law enforcement.

A situation also can occur during an interview in which a juvenile or family member discloses that they are intending to commit arson or another type of criminal act. If this occurs, the interviewer must take certain steps. First, the interviewer should establish whether there is an actual plan, with time and place identified, for committing the crime. If this is the case, then it is the interviewer's responsibility to follow a course of action to prevent the intended criminal activity. The interviewer also must inform those intending to commit the crime of this responsibility. The interviewer then must take the necessary steps to prevent the intended criminal act, including notification of the proper authorities as well as the intended victims. Disclosure of intention to commit arson or any other crime must be taken seriously and the interviewer must move to prevent the occurrence of the criminal activity.

Confidentiality

Juveniles and their families will want to know whether what is said during the interviews will be communicated to others. The nature of the relationship between the interviewer, the juvenile, and the family is one of reasonable trust. However, because juveniles and parents share information separately, they may want to share some of their information in confidence. If this request is made, the interviewer must clearly state that, while the confidence can be kept, if the interviewer determines that it is in the best interest of the juvenile or parent to disclose the information, they will do so. However, prior to sharing the confidence, the interviewer will inform the juvenile or parent that the confidence will be disclosed. The general rule is that confidences will be disclosed

Disclosure of intention to commit arson or any other crime must be taken seriously, and the interviewer must move to prevent the occurrence of criminal activity.

Table 3.5 Warning Signs of Abuse		
	PHYSICAL ABUSE	**SEXUAL ABUSE**
VICTIM	Unexplained Bruises	Genital Trauma
	Welts	Venereal Disease
	Bite Marks	Sleep Disturbances
	Burns	Bedwetting
	Fractures	Abdominal Pain
	Lacerations	Irregular Appetite
	Abdominal Injuries	Dramatic Weight Change
	Hair Loss	
	Upper Body Injuries	
PERPE- TRATOR	Temper Outbursts	Sophisticated Knowledge of Sex
	Destructive	Promiscuous
	Physically Violent	Prostitution

when the shared information is potentially harmful to those sharing it or to others. The decision of disclosure rests with the interviewer.

Juvenile firesetters and their families also will want to know whom and under what circumstances others outside the interview situation will have access to the verbal and written information collected during the interview. That is, can other individuals or agencies request and receive interview information regarding the juvenile and the family. There are several situations in which these requests might occur. In general, parents must give permission to disclose information collected during the interview. Also, there are certain legal situations, such as subpoenas of records, in which it is mandatory to hand over

the information. In addition, in many states child protective services has the right to request information. However, in some states there have been recent changes in juvenile code allowing information about the juvenile interview to be shared between agencies such as the fire service and juvenile justice, without parental permission. Chapter 6 reviews in detail the conditions of confidentiality regarding the verbal and written information collected during the interview.

Supplementary Interviews

While the majority of information about firesetting and related problems will come from the juvenile and family during their interviews, there are other sources of information that may be of value. If a fire incident report exists for the case, its content should be reviewed. If fire investigation conducted a separate interview, then this information may be important. Also, family friends often can provide useful information, and teachers and counselors at school can add important observations. These interviews can confirm and expand upon the information gathered from juvenile and family interviews. If these additional interviews are conducted, it is essential that the parents of the juvenile give their written permission. Examples of written release of information forms that parents can sign if they agree to supplemental interviews are included with both the Risk Surveys and the Comprehensive FireRisk Evaluation. These are general waivers or releases that allow other agencies or referral sources to obtain information from the interviews. There are reasons why parents may or may not want friends or school personnel involved in these interviews, and it is their privilege to make this decision. There is one circumstance in which parental permission is not necessary when contacting the school, family friends, or other community agencies. This is the circumstance where public safety may be in jeopardy, such as the threat of starting a fire. In this situation, fire officials, arson investigation, and law enforcement do not need parental permission to contact the relevant persons or agencies.

Legal Issues

There are several legal concerns that surface when conducting evaluation interviews. **Table 3.6** lists some major legal questions that most juvenile firesetter programs will need to address. Because every town, county, and state have different regulations and laws, the best way to answer these questions is for each juvenile firesetter program to seek advice from its local experts. These experts can be found in the local or state fire service and in related community agencies. Professionals, such as the city attorney, the local district attorney, or the juvenile court judge can offer expert advice on legal questions. Following the correct legal procedures will ensure the protection of not only the juvenile and the family, but the operation of the juvenile firesetter program.

Table 3.6
The Interview
Common Legal Questions

- Do you need permission to interview the firesetter?

- Do you need permission to interview school counselors and teachers?

- Do you need permission to interview family and friends?

- What kind of permission do you need to disclose verbal or written information collected during an interview? To interview participants? To speak to persons other than interview participants, such as mental health professionals and juvenile justice?

- Can the written records of the interview be subpoenaed by a court of law? Can the interviewer also be subpoenaed to testify in court?

- What kinds of action do you take when there is disclosure of a committed crime during an interview?

- What kinds of actions do you take when a threat is made to commit a crime during an interview?

The Juvenile Firesetter Child and Family Risk Surveys

In September 1995, the Colorado Department of Public Safety/Division of Fire Safety was awarded a federal grant to design and test the application and effectiveness of the Juvenile Firesetter/Arson Control and Prevention Program model for statewide dissemination. The Colorado project determined that the fire service needed a fire risk assessment instrument that was accurate in predicting future risk for firesetting, yet offered a reduction in the amount of time to conduct an evaluation interview. They designed a two step procedure to accomplish this objective. First, the Colorado group revised the evaluation instrument developed by Dr. Fineman, published in previous USFA Handbooks, and currently used in juvenile firesetter programs across the country. This revision not only formed the basis of their planned research, but it contributed significantly to the current version of the Comprehensive FireRisk Evaluation.

The second step was to develop a measure of firesetting risk by selecting the most statistically valid questions contained on the Comprehensive FireRisk Evaluation. This study was conducted by Moynihan and Flesher (1998) and is described in a research paper referenced in suggested readings at the end of this chapter. The result of this work was the Juvenile Firesetter Child and Family Risk Surveys. The Fire Risk Surveys consist of two parts, the Family Risk Survey and the Child Risk Survey.

The Family Risk Survey contains two sections. The first is an introductory section that records demographic information. The survey is comprised of seven questions, accompanied by scoring instructions. The scores on the survey are linked directly to specific recommendations for intervention. If parents are initially unable to come in for a personal interview, this Survey can be conducted by telephone. However, this is not recommended as routine practice, and the results of the telephone interview should be followed up by a personal interview with the parents and the juvenile.

The Child Risk Survey consists of five sections. The first section collects demographic information. The second section presents an informational activity for the child. The third section is the development of rapport. The fourth section is an interview exercise for children nine and over asking them to describe their most recent firesetting incident from beginning to end and then in reverse order. The fifth section contains fourteen questions, with scoring instructions. As with the Family Risk Survey, the scores on the Child Risk Survey are linked directly to specific recommendations for intervention.

The Child and Family Risk Surveys are contained in **Appendix 3.1**. Instructions for application and scoring also are included in the appendix. In addition, there are four items that are recommended to accompany the administration of the Risk Surveys. The first is a participation release signed by parents, that indicates they have an understanding of the evaluation procedure, give permission for their child to participate, and authorize release of information to other community agencies. The second item is a release of liability, also signed by parents, holding harmless the Juvenile Firesetter program. The third item is a release of information, signed by parents, which allows the Juvenile Firesetter Program to request information about their child from other community agencies. The final item is a risk advisement, signed by parents, that indicates they understand that their child has a serious risk of continued involvement in firesetting and that they have been advised to seek help from mental health. These items also are recommended for use with the Comprehensive FireRisk Evaluation.

The Comprehensive FireRisk Evaluation

The development of the Comprehensive FireRisk Evaluation has taken place over the last two decades under the direction of Dr. Kenneth Fineman. It has been published in previous USFA Handbooks, and many fire departments throughout the country have been trained to use this instrument to evaluate juvenile firesetters and their families. From an analysis of these evaluations, the instrument has undergone many applications and revisions. The suggested readings section at the end of this chapter lists references describing the development of this instrument. **Appendix 3.2** contains the most current version of the instrument. This appendix also includes all the necessary items to implement the instrument, including a participation release, a release of liability, a release of confidential information, and a risk advisement. Some juvenile firesetter programs have developed their own version of the Comprehensive FireRisk Evaluation, including the Fire Stop Program in Indianapolis. This version is referenced in suggested readings at the end of this chapter.

The Comprehensive FireRisk Evaluation consists of four parts. The first part is the Family FireRisk Evaluation and score sheet used for interviewing parents. The Family FireRisk Evaluation consists of one introductory section, and nine content areas. The introductory section contains demographic information regarding the child, parents, other family members, and the firesetting incident. The interviewer must complete all of this information. The nine content areas include health history, family structure issues, peer issues, school issues, behavior issues, fire history, crisis or trauma, characteristics of the firestart, and observations. Each content area contains questions that must be scored by the interviewer. The interviewer can select the manner in which the questions are asked, but information must be obtained from the parents to arrive at a score for each question.

The second part is the Juvenile FireRisk Evaluation used for interviewing youth. There is a score sheet that accompanies it. In addition to the Development of Rapport and Determine Level of Understanding sections, there are eight content areas. They are: school issues, peer issues, behavior issues, family issues, crisis or trauma, fire history, characteristics of the firestart, and observations. Each content area contains a specific set of questions. The interviewer must be able to collect the information and score the answers to all of these questions. In addition, the interviewer can add their own observations and written comments.

There are instructions that accompany the Juvenile FireRisk Evaluation describing how to score the interview.

The third part is the Parent FireRisk Questionnaire that is completed by parents in writing at the time of the interview by parents. While the juvenile is being interviewed, parents are asked to complete the Parent FireRisk Questionnaire. This questionnaire, in English and Spanish, is contained in **Appendix 3.2**. The instrument is divided into eight sections that include questions about the child's school, health and general development, peers, antisocial behavior, symptoms of anxiety or depression, fire history, family concerns, and severe behavior dysfunction. The Parent FireRisk Questionnaire uses the same scoring system as the Juvenile and Family FireRisk Evaluations. A Visual Key to scoring the Parent FireRisk Questionnaire is also presented in **Appendix 3.2**.

The final part is the Comprehensive FireRisk Analysis that summarizes the interview and questionnaire scores and calculates firesetting risk. The total scores from the Parent FireRisk Evaluation, the Juvenile FireRisk Evaluation, and the Parent FireRisk Questionnaire are entered into formulae for computing Juvenile, Family, and Total Risk percentages. These percentages indicate the level of risk- little, definite, or extreme- for the juvenile and family. These scores also estimate the likelihood that the juvenile and family will experience firesetting or other behavior problems in the future.

Summary Points

- **The purpose of evaluating juvenile firesetters and their families is to determine their risk for involvement in future firesetting.**

- **The outcome of an evaluation is the classification of juveniles and their families into three risk levels- little, definite, and extreme. Each risk level relates to a specific intervention.**

- **The structured interview is the method-of-choice for conducting an evaluation of juvenile firesetters and their families.**

- **There are six essential elements to consider when conducting an evaluation interview with juvenile firesetters and their families. These six elements include the target population, the interview format, the interview style, special situations, confidentiality, supplementary interviews, and legal issues.**

- **The Juvenile Firesetter Child and Family Risk Surveys offer a statistically validated method for conducting a brief evaluation interview with juvenile firesetters and their families. The surveys result in the classification of firesetting risk levels and recommendations for intervention and referral.**

- **The Comprehensive FireRisk Evaluation provides an extensive evaluation interview with juvenile firesetters and their families. It yields a classification of firesetting risk levels, information on psychological and social behaviors, and recommendations for intervention and referral.**

Suggested Readings

Topic	**Resource**
A research paper describing the development of the Juvenile Firesetter Child and Family Risk Surveys.	Moynihan, M. and Flesher, E. Locating a Risk Cut-Off Level Based on Key Variables in the Regression Equation. Child Interview. Parent Interview. Boulder, Colorado: Department of Psychology, University of Colorado, 1998.
The development of the Comprehensive FireRisk Evaluation.	Fineman, Kenneth. A Model For the Qualitative Interview of Child and Adult Fire Deviant Behavior. American Journal of Forensic Psychology, 1996, 13, 31-60.
This is an alternative version of the Comprehensive FireRisk Evaluation developed by the Fire Stop Program. It offers a detailed structured interview format for evaluating juvenile firesetters and their families.	Spurlin, Barbara. Fire Stop Program. Indianapolis Fire Department. Juvenile and Family Fire Risk Interview Forms (1997).
Advanced interview skills.	Lewis, M. Psychiatric Assessment of Infants, Children, and Adolescents. Child and Adolescent Psychiatry. In Melvin Lewis (ed). Child and Adolescent Psychiatry. A Comprehensive Textbook. Baltimore: Williams and Wilkins, 1996. pp 440-457.

Chapter 4

Intervention

The goal of a juvenile firesetter program is early identification and intervention of at-risk youth. A juvenile firesetter program is one part of a network of community services designed to prevent and control firesetting and arson-related activities. There are five essential components of a juvenile firesetter program--identification to assignment, evaluation, education, referral, and exit to follow-up. Additional components capable of enhancing the operation of a juvenile firesetter program are community service, restitution, and counseling. There are a number of excellent examples of successful juvenile firesetter programs.

Objectives

4.1. **To describe a continuum of care to help reduce juvenile involvement in firesetting and arson.**

4.2. **To define the five essential components of a juvenile firesetter program.**

4.3. **To identify additional components to enhance a juvenile firesetter program.**

4.4. **To describe the components of seven successful juvenile firesetter programs.**

Continuum of Care

Communities must take the lead in building a comprehensive strategy to combat juvenile firesetting and arson. The centerpiece of this approach is a continuum of care designed to provide swift, certain, and consistent intervention for all youth.

A continuum of care system provides a range of interventions. This range of interventions is based on corresponding levels of fire behavior and classification of firesetting risk. **Table 4.1** presents the continuum of care. It illustrates the relationship between levels of fire behavior, classification of firesetting risk, and points of intervention.

On the continuum of care, each level of fire behavior has a corresponding classification of firesetting risk and point of intervention. For example, a youngster engaging in unsupervised firestarts, that is lighting matches the first time without parental knowledge, must be identified as soon as possible. Parents, caregivers, or the fire service usually are the first to identify the problem. Once referred to a juvenile firesetter program, they can be assessed, classified according to risk, and provided the appropriate type of intervention. In the case of unsupervised firestarting, the youngster is likely to be classified as little risk and provided an educational intervention designed to eliminate future involvement in unsafe firestarts.

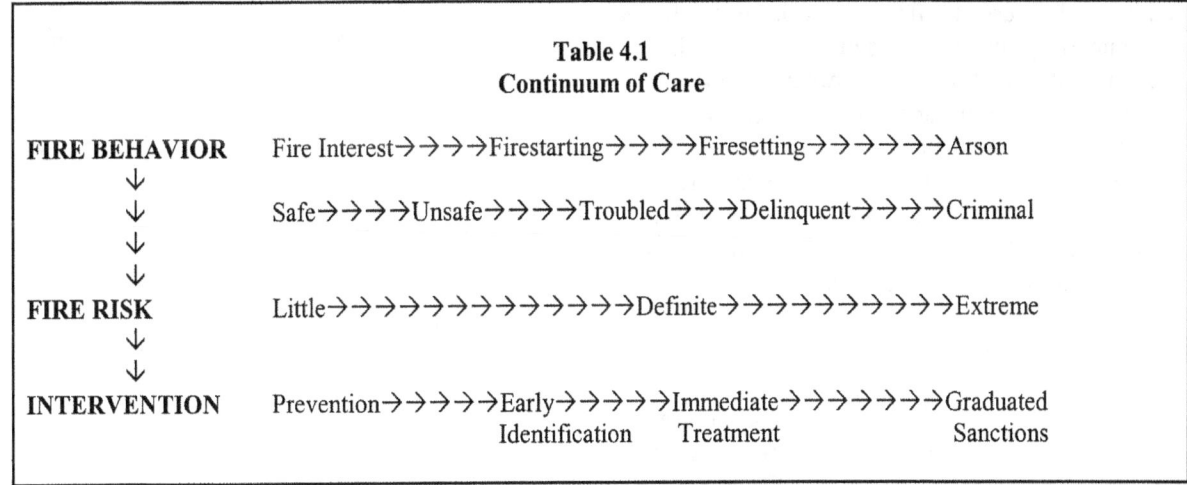

Table 4.1
Continuum of Care

FIRE BEHAVIOR Fire Interest→→→→Firestarting→→→→Firesetting→→→→→→Arson
  ↓
  ↓   Safe→→→→Unsafe→→→→Troubled→→→Delinquent→→→→Criminal
  ↓
  ↓
FIRE RISK   Little→→→→→→→→→→→→→Definite→→→→→→→→→→→Extreme
  ↓
  ↓
INTERVENTION Prevention→→→→→Early→→→→→Immediate→→→→→→Graduated
                Identification   Treatment   Sanctions

Points of Intervention

There are four major points of intervention on the continuum of care. They are prevention, early identification, immediate treatment, and graduated sanctions. Prevention efforts teach all children of all ages fire safety and survival skills. Early identification classifies at-risk youth and recommends specific strategies to eliminate future involvement in firesetting. Immediate treatment offers rapid access to mental health, social services, or other types of needed therapy. Graduated sanctions, using restrictive interventions, provides rehabilitation and correction services to juveniles involved in repeated firesetting and arson-related activities. Each of these four points represent increasingly intensive methods of intervention corresponding to increasingly severe levels of firesetting behavior. These four points of intervention also relate to levels of community programs and services that help reduce juvenile firesetting and arson. These programs and services are the topic of Chapter 5.

An effective juvenile firesetter program will have linkages to all four points of intervention on the continuum of care. However, the primary function served by a juvenile firesetter program is that of early identification. A juvenile firesetter program is in the unique position of identifying at-risk youth because it is the community agency most likely to be contacted when parents, caregivers, schools or others recognize children with a fire problem. In addition, because the fire service is most often the site of a juvenile firesetter program, there can be strong linkages between a juvenile firesetter program and the suppression or investigation units that first respond to fire scenes where youth may be involved. Early identification of at-risk juveniles is critical to swift and effective intervention aimed at eliminating future firesetting incidents.

The Juvenile Firesetter Program

The primary goal of a juvenile firesetter program is early identification and intervention to prevent and control firesetting and arson. Communities can build their juvenile firesetter program based on a number of different factors, including assessed needs, capabilities, and available resources. These program planning and development decisions are covered in detail in Chapter 6. It is recommended that communities consider five essential components when building their juvenile firesetter program. They are identification to assignment, evaluation, education, referral, and exit to follow-up. Together, these five components comprise the integrated structure of a juvenile firesetter program. **Table 4.2** illustrates the relationship between the five essential components of a juvenile firesetter program.

1. Identification to Assignment

The point of entry into a juvenile firesetter program is the identification of at-risk youth. There are a number of ways juveniles are identified. Parents, upon finding unspent matches in their child's pant pockets and small burn marks on toys and furniture, call their local fire department for advice. A school experiencing a series of trash can fires, identify one or more youths involved in the incidents, and contact the fire service for help. Fire suppression, called to a second story house fire, discover that a teenage boy left his five year old brother unattended for an hour only to return to find him playing with a lighter. These juveniles clearly are having problems with fire and need some type of intervention. A juvenile firesetter program can be their point of first contact.

Table 4.2
The Five Components of a Juvenile Firesetter Program

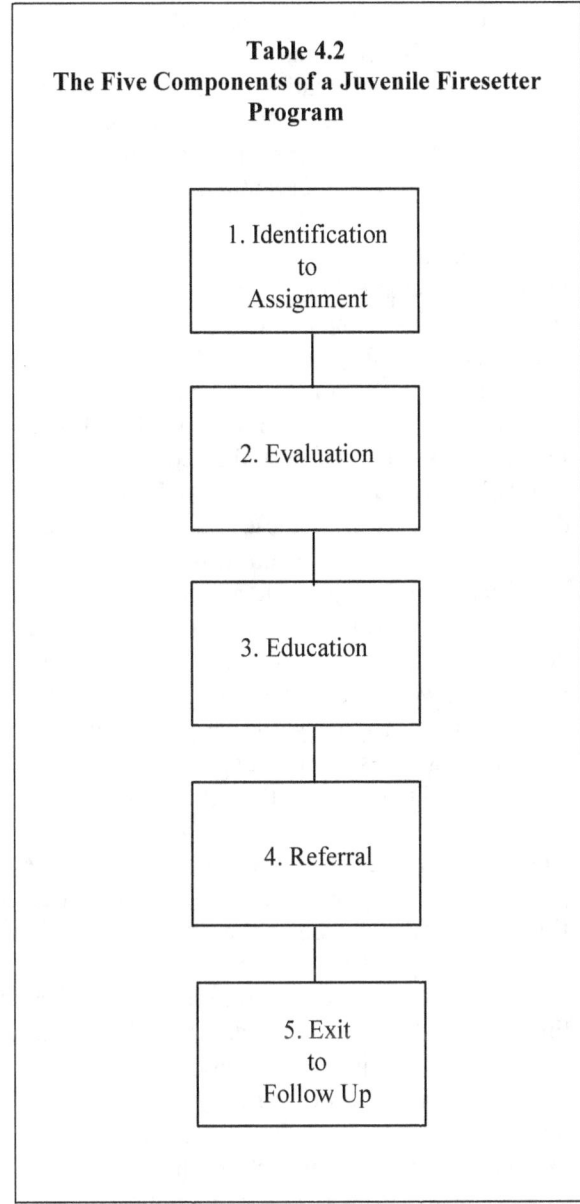

1. Identification to Assignment

2. Evaluation

3. Education

4. Referral

5. Exit to Follow Up

Early Identification

The earlier the identification is initiated the better the chances of a successful intervention. Early is defined in several ways. First, early is defined by the age of the child and by the absence of a firesetting history. For example, a young child involved in a first-time, unsupervised firestart motivated by curiosity or experimentation can be helped easily and effectively. The older child, involved in repeated firesetting, will be more difficult to help. The longer firesetting persists, the more difficult it is to correct. This is why it is critical

that at-risk juveniles be identified and offered immediate intervention.

Secondly, early can be defined as the identification of emerging psychological or social conflicts contributing to the firesetting behavior. It is not just the young child motivated by curiosity or experimentation that can be identified and helped by a juvenile firesetter program. Those juveniles involved in repeated firesetting incidents motivated by underlying psychological or social conflicts can be identified, evaluated, and referred to appropriate types of treatment. The earlier in this behavioral cycle these youth are identified, the more likely they will be helped to overcome their problems.

Finally, early can be defined as interrupting the beginning pattern of antisocial behavior that can lead to later criminal activity. Those juveniles setting fires motivated by malicious intention need to be identified early in their criminal careers. The earlier they are identified and helped, the better their chances of redirecting their lives. The more firmly established a criminal behavior pattern, such as repeated firesetting, the less effective sanctions or rehabilitation efforts are likely to be in preventing that behavior.

Typically, parents, community agencies, and fire suppression or investigation are points of initial contact for at-risk youth entering a juvenile firesetter program. If a fire department is the site of a juvenile firesetter program, the local fire stations may experience walk-in cases from the neighborhood. A juvenile firesetter program must have a set of procedures in-place to receive cases. These procedures must include the designation of specific intake or contact persons, a system for recording incoming cases, and specific methods for communication and referral. Chapter 2 describes in detail the identification procedures for a juvenile firesetter program.

Assessment

Once the juvenile firesetter program has identified at-risk youth, the next step is an assessment action. Assessment actions for juveniles can be legal and or voluntary, depending on the presenting firesetting incident. Chapter 2 describes these assessment actions. A juvenile firesetter program typically accepts youth and their families who voluntarily seek help, although there are situations in which juveniles can be mandated to participate in the program. These circumstances are described in Chapter 2. Immediately following an assessment decision, assignment of the juvenile and family to the program is initiated.

Assignment

An effective assignment to the juvenile firesetter program involves systematic screening procedures and a formal intake process. These functions are described in Chapter 2. Screening assures swift and immediate entry into the program. The intake process is designed to provide a secure and standard pathway through the juvenile firesetter program for all juvenile firesetters and their families.

Table 4.3 illustrates the relationship between component one (Identification, Assessment, and Assignment) and component two (Evaluation) of the juvenile firesetter program. In both components, decisions must be reached regarding the status of the juvenile and family. In **Table 4.3** these decisions-points are illustrated with circles and branching options. In component one, there are assessment decisions regarding the legal versus voluntary status of the juvenile and assignment decisions involving screening and intake. Once these decision pathways are complete, then the juvenile and family can proceed to the second component of the program - Evaluation. In Evaluation, a decision is made regarding the severity of the firesetting problem.

2. Evaluation

The main objective of the evaluation component of a juvenile firesetter program is to understand why children set fires. An evaluation of juvenile firesetters and their families involves a determination of the nature and severity of the presenting firesetting behavior. The purpose of

evaluation is to determine the likelihood or risk that another firesetting incident will occur in the future. Risk determination classifies firesetting into three levels - little, definite, and extreme - and provides the basis for recommending appropriate types of intervention.

A structured interview is the recommended method for evaluating juvenile firesetters and their families and arriving at a risk determination. Chapter 3 presents and describes two instruments for conducting structured interviews, the Juvenile Firesetter Child and Family Risk Surveys and the Comprehensive FireRisk Evaluation. The similarities and differences between these instruments are outlined in Chapter 3. Both instruments yield a classification of the juvenile and family into one of the three levels of firesetting risk. The decision of which instrument to use rests entirely with the juvenile firesetter program and will depend on the program's service goals, available resources, and desired outcomes. These issues are discussed in more detail in Chapter 6.

Once the evaluation procedure classifies the juvenile and the family into one of the three firesetting risk levels, the next step is recommending the appropriate intervention strategy. For example, all juveniles classified as little risk will be recommended for educational intervention. While definite risk juveniles may benefit from education, nearly all of these juveniles with be referred for immediate treatment such as mental health or social services. Those classified as extreme risk are unlikely to benefit from education, and must be referred either to immediate treatment or to some type of sanction that may include juvenile justice intervention. **Table 4.4** presents the risk determination levels (little, definite, and extreme) and corresponding intervention strategies (education and referral).

3. Education

At least six in ten cases identified by a juvenile firesetter program will be classified as little risk. That is, accident, curiosity, or experimentation motivates the firesetting of these juveniles. The recommended intervention strategy for these

Table 4.3
Identification and Evaluation

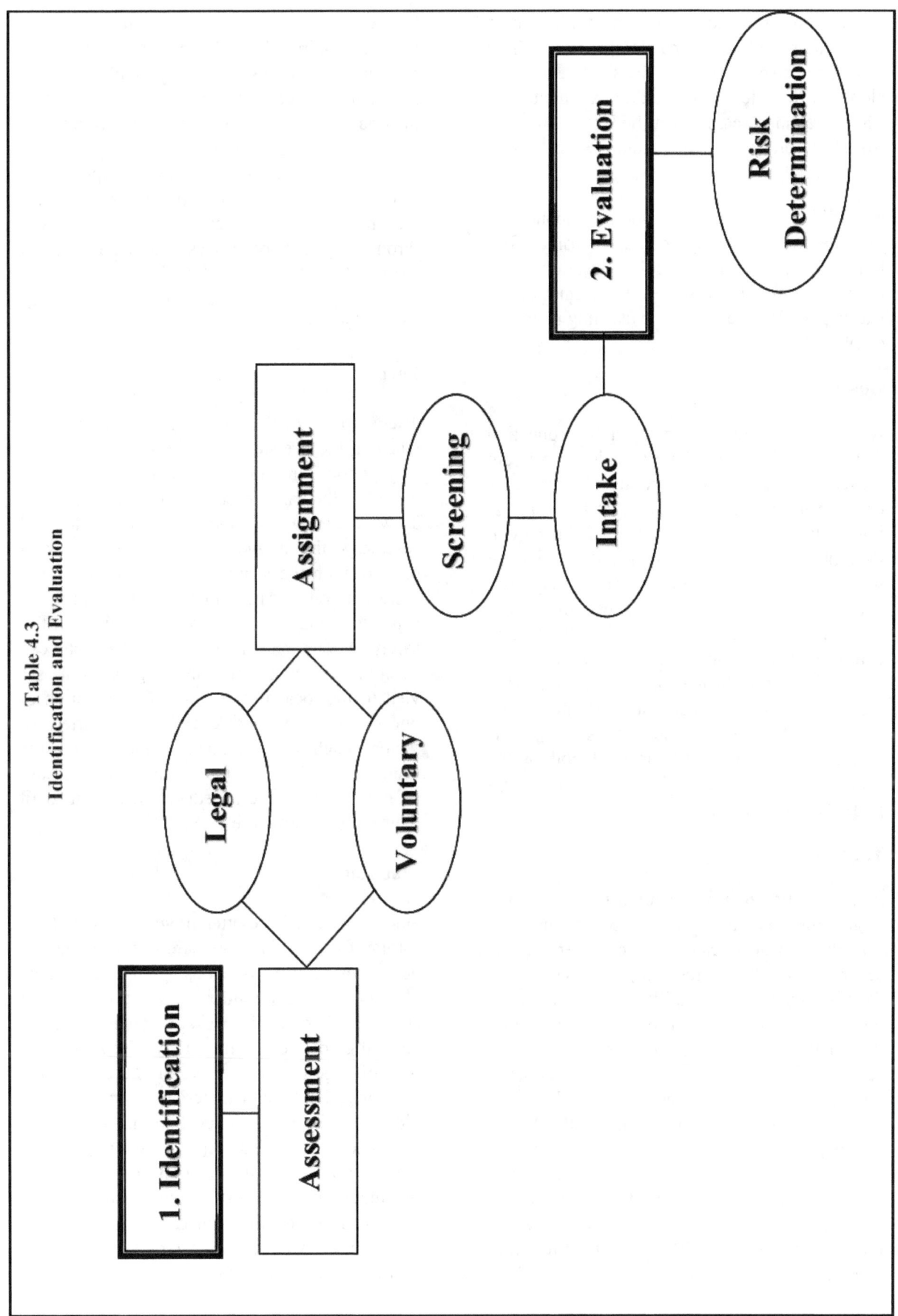

cases is fire safety education. In addition, three in ten cases are likely to be classified as definite risk. They also will require fire safety education along with referral to additional treatment. Therefore, most cases identified by a juvenile firesetter program will require fire safety education.

To build a fire safety education component, a juvenile firesetter program must consider four important factors. They are the education goals, the target group to be served, the format of the learning environment, and the teaching materials employed.

Goals

A juvenile firesetter program's education component has two primary goals. They are prevention and survival. A juvenile firesetter program must offer fire prevention and survival information to both the youth and the parent. Prevention education should keep the juvenile who is curious about fire or who perhaps has made a wrong choice about firestarting, from using fire as a destructive tool. Prevention education also should raise the awareness of parents regarding their own behavior and what they can do to create a fire-safe home environment. Survival education should focus on those skills that must be learned, both by the juvenile and parent, to help save lives and property in case of fire.

Target Group

A juvenile firesetter program must determine the target group served by the education component. The three critical features to consider are the severity of the firesetting behavior, the developmental level or ability of the juvenile to understand and learn fire safety education information, and the age of the youth. The severity of the firesetting behavior is determined during evaluation by the assigned risk level. A fire safety education component should be aimed primarily at the little risk level, with a secondary emphasis on definite risk. The developmental level or ability to understand and learn the educational information is related to the age of the juvenile. Therefore, the fire safety materials employed should be age-appropriate.

For example, if some educational materials require reading, it is important to know the reading capabilities of the juvenile. Many programs target their fire safety education materials for the following age categories: preschool and kindergarten or six years and under, grades one through three or ages seven through ten, grades four through six or ages eleven through thirteen, and grades seven through high school or ages fourteen and up. A section to follow will describe recommended fire safety resources according to these age categories.

Format

There are a number of different formats for teaching the fire safety education component of a juvenile firesetter program. **Table 4.5** describes the various teaching formats. There is a wide range of options, from individual sessions with the juvenile and parent to working directly with the schools. Many programs use more than one format. For example, a program can offer young children not only individual fire safety lessons, but these fire safety lessons can be combined with family meetings as well as a visit to the local fire station. The selection of one or more formats depends on a variety of factors such as available resources and the training of staff. Each juvenile firesetter program can choose an education format to fit their program structure.

Materials

Once a juvenile firesetter program selects their format for teaching fire safety education, the next consideration is choosing the educational materials. **Appendix 4.1** presents an educational materials package containing some general guidelines for fire safety education and a list of age-appropriate resources, including programs, books and materials, brochures, and videos. Bilingual education materials are indicated in the listing. **Appendix 4.1** also notes how to obtain this information and the cost of the materials. It must be emphasized that a juvenile firesetter program can obtain many of their fire safety education materials at little or no cost from various resources.

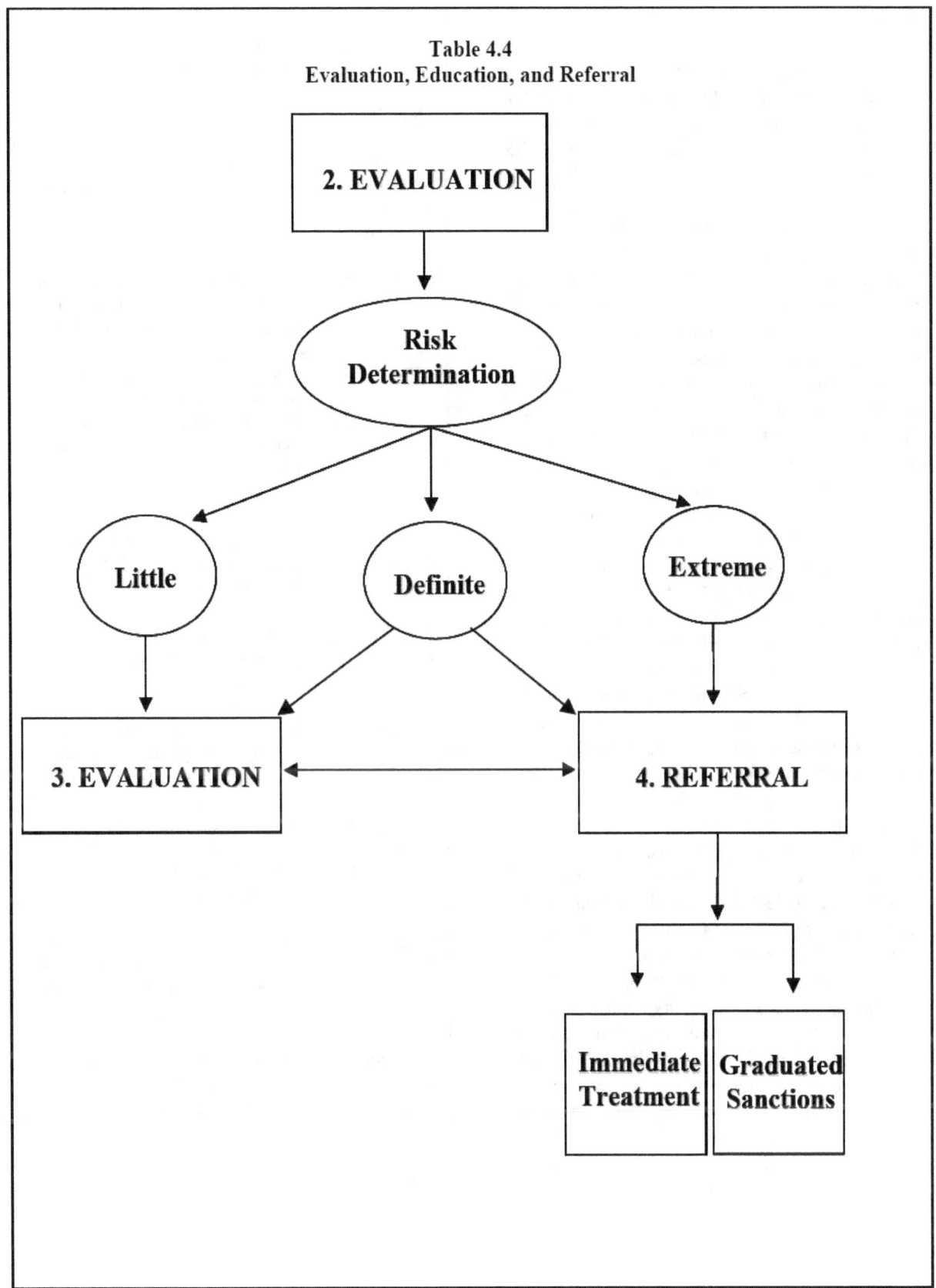

Table 4.4
Evaluation, Education, and Referral

2. EVALUATION

Risk Determination

Little

Definite

Extreme

3. EVALUATION

4. REFERRAL

Immediate Treatment

Graduated Sanctions

4. Referral

The identification of definite and extreme risk firesetters necessitates that a juvenile firesetter program has in-place an effective referral network. Definite and extreme risk juvenile firesetters need access to a variety of intervention strategies not offered by a juvenile firesetter program. These services can include mental health, law enforcement, social services, and juvenile justice. A juvenile firesetter program must convince parents of definite and extreme risk youth to follow through with their referral. In many cases, their referral is to mental health services. When the referral is voluntary, it is necessary that parents understand the importance of obtaining the recommended help. One program asks parents to sign a document stating that they understand that their child is at-risk for future involvement in firesetting and that it is necessary to receive additional help to resolve the problem. (Please see either **Appendix 3.1 or 3.2** for the Risk Advisement document.) When the referral is court-ordered, if follow through does not occur, the juvenile justice system has the authority to impose harsher restrictions or sanctions on the youth. To avoid this undesirable alternative, families choose to comply with the intervention recommended by the juvenile firesetter program.

A juvenile firesetter program can provide the linkages to community services by setting-up referral pathways. The four major steps to building an effective referral system for a juvenile firesetter program are identifying the network of services, contacting the service agencies, developing the referral agreement, and maintaining a method of quality control for these services. These topics are covered in detail in Chapter 5. **Table 4.6** presents the scope of the referral network. A juvenile firesetter program must have linkages to all of these community services. When a definite or extreme risk firesetter is identified, it is the obligation of the juvenile firesetter program to refer these cases to the appropriate community agency.

Table 4.5
Teaching Formats

Method	Description
Telephone Interviews/ Mailed Materials	A prepared response to calls requesting information and educational materials are mailed to the home.
Home Visits	Upon request, an assessment is conducted and educational material is provided in the home.
Office Visits	One to three visits for the purpose of assessment, evaluation, and education. Educational material is provided and homework between visits is typical.
Safety Lessons	A variety of fire safety education projects to be completed by the child and parents at home.
Family Safety Meetings	Special fire safety education projects encourage family discussion.
Station Visits	Young children with their parents are given educational tours of their local fire station.
Station Work Time	Older children spend supervised time at the fire station learning some of the responsibilities of being a firefighter.
Clinics	Regularly scheduled group meetings for parents and separately for juveniles to teach fire safety education.
Working with the Schools	The fire service works directly with the schools to coordinate a fire education safety program.
Teaching Aids and Resources	Audio-visual materials are effective methods for teaching fire safety education.

5. Exit to Follow-Up

There are several points at which youth can exit a juvenile firesetter program. They can exit voluntarily at any time. Some juveniles may exit after identification and assessment or after evaluation, depending on the nature and severity of the presenting problem. However, the typical exit for most juveniles is after education or referral. There are two types of exits. There are those cases that exit without referral to other community agencies and those cases that exit with referral to additional services. It is important that follow-up procedures are set-up for both types of exits so that youth and their families understand that the juvenile firesetter program will continue to be concerned about their welfare.

The two important features of follow-up procedures are when and how they are implemented by the juvenile firesetter program. For all cases, a primary follow-up is recommended four to six weeks after exit. If resources are available, a secondary follow-up can take place between six to twelve months after exit.

A juvenile firesetter program can conduct follow-up procedures in a number of different ways. Follow-up methods include telephone calls, written contacts, and visits. Telephone calls are the most cost-effective and least time consuming method of follow-up. Written contact can include postcards, letters, surveys, and electronic communication. Return visits require the most resources, but allow for a direct assessment of the firesetting problem. In addition to selecting the method of follow-up, a juvenile firesetter program must consider the content of the follow-up contact. The content can be a standard set of questions for all cases or a set of questions designed specifically for each case. **Appendix 4.2** presents sample follow-up surveys that can be used either as a telephone contact or a written communication. Follow-up procedures not only help to reinforce fire-safe behavior for juveniles and their families, but they also provide information on the program's effectiveness in reducing involvement in firesetting and arson.

Additional Components

There are three additional components that programs can use in their work with juveniles. They are community service, restitution, and counseling. These components can enhance the operation of a juvenile firesetter program, but are not considered to be critical for its success.

Community Service

Community service typically is employed with older juveniles who have engaged in repeated firesetting incidents. It is used as a consequence of firesetting, and often is accompanied by additional sanctions. There are a variety of community service activities such as food and clothing drives, senior citizen assistance programs, clean-up efforts at parks and beaches, and station work-time. Station work-time is an effective form of community service used by several fire departments. It involves spending time at the fire station learning what it means to be a firefighter. Some of the supervised activities include putting a vehicle back in service, assisting in washing and cleaning the vehicle, wearing bunker gear, and discussing with firefighters the dangers of the job and how they feel after a call. This first-hand experience leaves a lasting impression on youth.

Restitution

Many states require juvenile and parent responsibility for the dollar damage caused by firesetting. For example, the state of Maryland has enacted two laws. The first requires restitution for the first $500 in such damages caused by fires set by juveniles, and the second holds juveniles responsible for their firesetting. Other states, through the diversion or probation departments of their juvenile justice system, have restitution programs. There are some juvenile firesetter programs that also implement restitution programs holding juveniles responsible for some or all of the dollar damage resulting from their firesetting. **Appendix 4.3** presents a restitution agreement used by the state of Colorado in their Juvenile Firesetter Prevention Program. Restitution is an intervention method that strongly emphasizes to

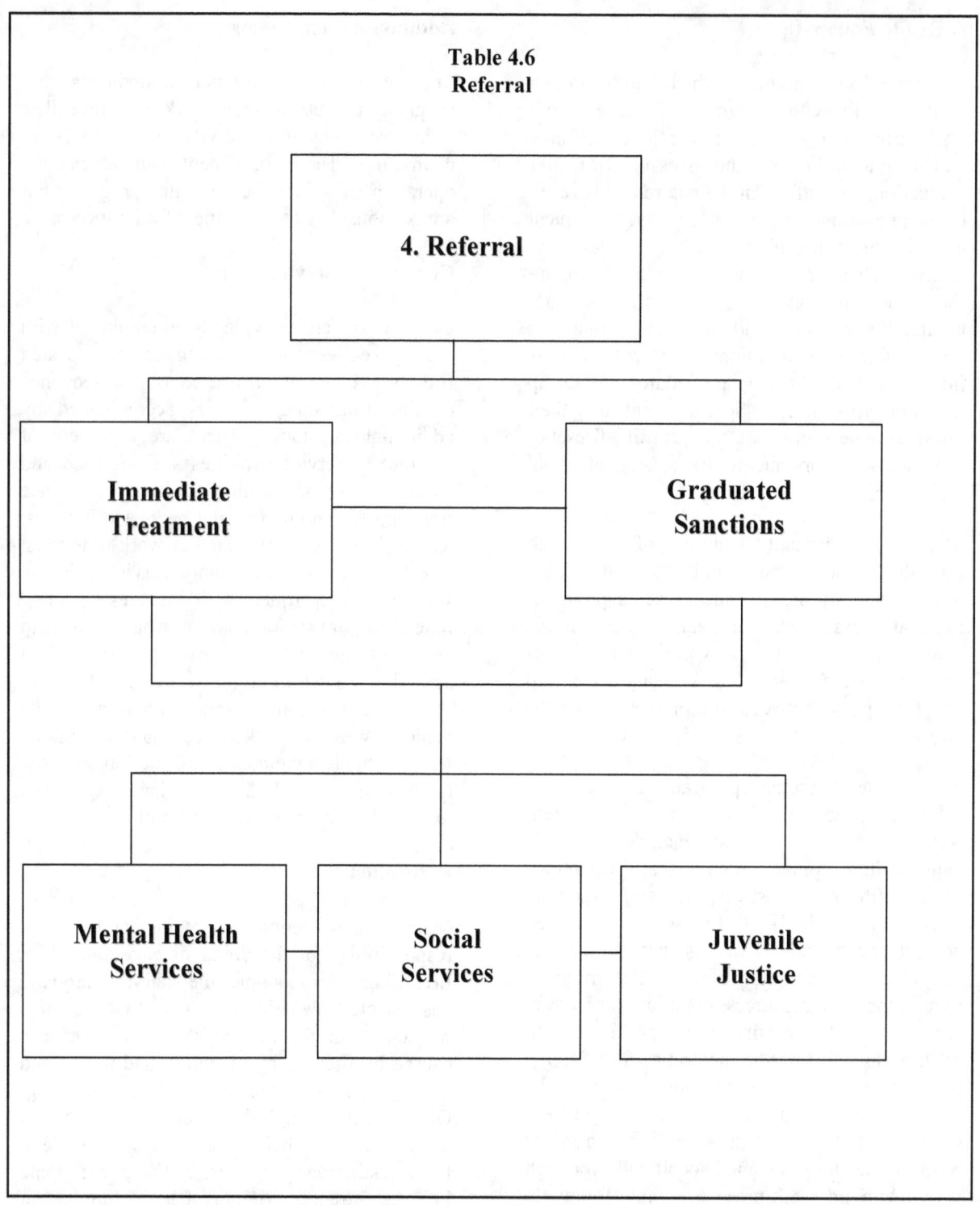

Table 4.6
Referral

4. Referral

**Immediate
Treatment**

**Graduated
Sanctions**

**Mental Health
Services**

**Social
Services**

**Juvenile
Justice**

juveniles and parents the consequences of firesetting.

Counseling

Certain juvenile firesetter programs employ counseling services as one component of their intervention strategy. For example, in King County, located near Seattle, Washington, the juvenile firesetter program uses a community teamwork approach where fire education specialists work with mental health professionals to create a county-wide referral network for juvenile firesetters. In addition to identification, assessment, evaluation, and education, mental health professionals offer up to four counseling sessions for juveniles and their families. If families cannot afford to fund these services, the program will provide financial assistance.

Another example of a juvenile firesetter program with a strong mental health component is the Phoenix Fire Department's Youth Firesetter Intervention Program. Recognizing the need for behavioral health services for firesetters and their families, they developed a panel of degreed and licensed mental health care providers. This panel receives extensive and continuous firesetter education training as well as practical experience working with firesetters and their families. The panel meets monthly for educational presentations and a review of their clinical cases. If mental health services are recommended during the assessment phase of the program, a health care provider will call the family within 48 hours to set-up an appointment. If the family cannot support these services through their insurance, funds from the Phoenix Fire Department will be used to assist them. This comprehensive plan provides an intensive and integrated intervention strategy for addressing the firesetting problem.

Juvenile Firesetter Programs--A Selected Sample

A large number of juvenile firesetter programs are operating successfully throughout the country. The purpose here is to illustrate a representative sample of these programs. They are selected based on their geographic distribution, the diversity of their program components, their ability to build and sustain program operations, and their existing documentation. References for these programs are found in the suggested readings section at the end of this chapter.

Seven juvenile firesetter programs comprise the sample. **Table 4.7** presents a summary description of their five program components-- identification to assignment, evaluation, education, referral, and exit to follow-up. If the programs use additional methods of intervention, they are also noted. The presentation of the material in this table is intended to provide an overview of available intervention options for developing and maintaining an effective juvenile firesetter program.

Prevention education should raise the awareness of parents regarding their own behavior and what they can do to create a firesafe home environment.

Summary Points

- **The four points of intervention--pre-vention, early identification, immediate treatment, and graduated sanctions--on the continuum of care represent a range of strategies designed to reduce juvenile involvement in firesetting and arson.**

- **A juvenile firesetter program is an early identification approach with strong linkages to all other points of intervention on the continuum of care.**

- **The identification to assignment and evaluation components of a juvenile firesetter program result in a risk determination and recommended set of intervention strategies.**

- Cases classified as little and definite risk by a juvenile firesetter program will benefit from fire safety education.

- All definite and extreme risk firesetters and their families will utilize the referral component of a juvenile firesetter program.

- Follow-up efforts by a juvenile firesetter program will let the juvenile and family know that there is a continued concern for their welfare.

- Community service, restitution, and counseling represent three additional services that can enhance the operation of a juvenile firesetter program.

Table 4.7
Examples of Juvenile Firesetter Programs

PROGRAM LOCATION	GENERAL DESCRIPTION	COMPONENTS				
		IDENTIFICATION TO ASSIGNMENT	EVALUATION	EDUCATION	REFERRAL	EXIT TO FOLLOW-UP
Fire Stoppers • King County Fire District Prevention Division 10828 S.E. 176th St. Renton, WA 98055 • (425) 255-0931	A comprehensive regional program designed to reduce the incidence of fireplay and firesetting through early identification, evaluation, education, and treatment. The success of this program is attributed to its strong coalition of representatives from law enforcement, juvenile justice, fire service, mental health, business/nonprofits, schools, and media.	The primary sources of referrals are caregivers, school staff, administrators or security, mental health, fire service, law enforcement, and juvenile justice. Contact is made with the caregiver of the youth involved in the firesetting incident. This may be done at the actual fire scene or by telephone shortly after the incident. A letter also may be sent in the event telephone contact does not take place.	The parent/caregiver accompanies the youth to an evaluation interview. Interviews are usually provided at (but not limited to) the fire station, the youth's home or school. A rights and responsibility release form is presented and signed. During the interview, demographic information is collected and a structured interview is conducted with the youth and parents. Interview forms are completed for the youth and parents.	Fire safety lessons, including both fire prevention and fire survival skills, are taught to the youth and parents. A variety of teaching techniques and materials are used to achieve the best result with each family. The Family's Response to Firesetting (a booklet) is provided to the parents. Fire safety contracts also may be introduced.	The interview forms are scored and a determination is made as to whether to refer the youth and family to mental health or to continue with educational efforts. The program provides eight hours of service for families referred to mental health.	Follow-up calls are made to obtain feedback on the program service and to track the progress of its clients. A follow-up form is completed three months after the education-only intervention. A follow-up occurs at 3 and 6 months once the youth and family have completed mental health intervention.

Table 4.7
Examples of Juvenile Firesetter Programs (Continued)

PROGRAM LOCATION	GENERAL DESCRIPTION	COMPONENTS				
		IDENTIFICATION TO ASSIGNMENT	EVALUATION	EDUCATION	REFERRAL	EXIT TO FOLLOW-UP
Fireproof Children • Rochester Fire Department and Fireproof Children 20 North Main Street Pittsford, NY 14534 • (716) 264-0840	This program consists of two approaches, prevention through education and the systematic follow through of all fires involving children. The fire department takes the responsibility of identifying the indicators of juvenile involvement in fireplay and firesetting, evaluates and classifies at-risk youth and their families, delivers educational intervention, and refers youth and families to additional intervention services as needed.	Youth and families can enter the program on referral from parents, community agencies, and fire investigation. Once identified, all youth and family are assessed. Interviews are conducted, and the family and home environment are checked. A preliminary decision is reached regarding further action. These decisions include, but are not limited to, legal action, evaluation, education, and referral.	There are four major goals of evaluating firesetting youth and their families. They are to acquire or confirm specific information, to assess motivation, to ascertain prior involvement in firesetting, and to describe the family, school, and social environment. The evaluation procedures outlined in this handbook are recommended.	This program advocates a two-part educational intervention involving both the fire department and the schools. Their educational approach utilizes the following teaching goals: the power of a match; avoid scare tactics and frightening examples; teach the responsible use of fire, and teach through the use of examples and age-appropriate hands-on experiences.	While the fire service is the lead agency providing evaluation and education services, a community network of cooperating agencies is available for referral. These agencies include law enforcement, criminal justice, social services, mental health, schools, and organized youth programs.	All youth and families identified by the program are followed up. The specific follow-up procedures depend on the types of recommended intervention. Reports are written and filed on all cases. Follow-up telephone calls, letters, and interviews also are employed. In addition, parents are encouraged to call the minute they detect any further fireplay or firesetting activity.

Table 4.7
Examples of Juvenile Firesetter Programs (Continued)

PROGRAM LOCATION	GENERAL DESCRIPTION	COMPONENTS				
		IDENTIFICATION TO ASSIGNMENT	EVALUATION	EDUCATION	REFERRAL	EXIT TO FOLLOW-UP
Juvenile Firesetter Intervention and Prevention Program • Saint Paul Department of Fire and Safety Services Division of Fire Prevention 100 East 11th St. Saint Paul, MN 55101 • (651) 228-6230 Fax: (651) 228-6241	This program is conducted in five phases. They are intake and assessment, education and referral, review, reinforcement and exit, follow-up and evaluation, and prevention. In addition, there are three separate program strategies aimed at three distinct age levels. The fireplay intervention programs focuses on ages 3-9, the juvenile firesetting program aims at ages 10-17, and the false alarm intervention program targets ages 3-17.	Referral to the program comes from fire suppression (70%), law enforcement (20%) and parents, schools and other community agencies (10%). All referrals are filtered through the public education officer. Fire and law enforcement refer to the program by completing a written form and parents and outside community agencies are typically referred by telephone or by letter.	A firesetter telephone risk survey is conducted for all referrals. All referrals are then scheduled to attend a two-visit program where additional evaluation material is obtained directly from the parent and child. In addition, written reports from fire investigation and law enforcement are reviewed. Parents also complete the parental information form.	This program has a very strong education phase. It is structured by age categories: Preschool-Kindergarten, Grades 1-3, Grades 4-6, and Grades 7-12. There are outcome objectives for each level. The education phase consists of two visits during which parent education and age-appropriate materials are used to teach fire safety. Homework assignments occur between visits. If the two education visits are successfully completed, the family is awarded a document of attendance.	If the initial assessment indicates the need for referral, the steps are taken. If mental health referral occurs, parents are given several options, including no cost, brief evaluation and counseling for children under 10. There are a number of other referrals for older youth and their families available on a sliding scale or at little or no cost, depending on the specific case and circumstance.	If education and or referral to mental health is completed successfully, a documentation of attendance or other official paper trail indicating that intervention has occurred is available for each case. A follow-up is conducted at one month, six months, and twelve months to assess firesetting behavior and whether the youth has demonstrated, verbally or by action, that safe behavior has been adopted. If firesetting has recurred, the youth is re-enrolled in the program and further steps are taken to secure counseling or other services.

Table 4.7
Examples of Juvenile Firesetter Programs (Continued)

PROGRAM LOCATION	GENERAL DESCRIPTION	COMPONENTS					EXIT TO FOLLOW-UP
		IDENTIFICATION TO ASSIGNMENT	EVALUATION	EDUCATION	REFERRAL		
Juvenile Firesetter Prevention Program • Parker Fire Protection District and Miller Safety Center 10795 South Pine Drive Parker, CO 80138 cpoage@parkerfire. org (303) 841-2608 • Colorado Depart. of Public Safety Juvenile Firesetter Prevention Program 700 Kipling Street Suite 1000 Denver, CO 80215 • (303) 239-5704	The development of this program began in Colorado's 18th Judicial District using the program components of identification, assessment, intervention, and follow-up. The success of this program provided an excellent base from which to test the statewide application of the juvenile firesetter intervention model in Colorado. Training programs were developed and implemented for the remaining 21 judicial districts in Colorado.	The program emphasizes the immediate documentation of all known cases of fireplay and fire setting. When a case is identified, a data base tracking system is established. A juvenile firesetter contact form is completed for each case, and procedures for storage, confidentiality, and sharing of information are developed. All fireplay and firesetting cases are assessed for risk.	There are two major strategies used to evaluate firesetting youth and their families. The individual program can select one or both options. The first option is the use of the Juvenile Firesetter Parent and Child Risk Surveys and the second option is the use of the Comprehensive FireRisk Evaluation. These procedures are described in detail in this handbook.	In the manual, Juvenile Firesetter Prevention Program, Volume 1, education intervention is described in detail. The emphasis is on teaching fire safety education. The program is presented in age categories: Three years and under, four to five years, five to seven years, seven to nine years, and ten to eighteen years. Each age category is comprised of general goals and teaching strategies, accompanied by specific fire prevention and survival activities.	Referral procedures are outlined in the manual, Juvenile Firesetter Prevention Program, Volume 1. Volume 2 in this series provides information for mental health professionals on treating juvenile firesetters and their families. Those cases classified as definite and extreme firesetting risk are referred to mental health or other necessary and appropriate interventions.		To evaluate the impact of the program, a follow-up procedure is required for all programs. Upon completion of the intervention, youth and families can be contacted by phone, mail, or in person. They are asked a series of questions to determine whether there has been any further involvement in fireplay or firesetting.

Table 4.7
Examples of Juvenile Firesetter Programs (Continued)

PROGRAM LOCATION	GENERAL DESCRIPTION	IDENTIFICATION TO ASSIGNMENT	EVALUATION	COMPONENTS			
				EDUCATION	REFERRAL	EXIT TO FOLLOW-UP	
Partners **A Juvenile Firesetter Intervention Program for the Community** • Palm Beach County Fire-Rescue 50 Military Trail Suite 101 West Palm Beach, FL 33415-3198 • (561) 233-0100 ext 331	This program is designed for the curiosity firesetter, the child influenced by peer pressure, or the child who just makes a wrong decision and ends up involved in a fire play incident. The primary goal of this program is to address children's curiosity about fires through a series of fire safety lessons and family safety meetings. This is an educational intervention program teaching children that a mistake was made and that it can be turned into a learning experience.	Youth can enter this program voluntarily or by court order. They can be referred by parents, caregivers, schools, the fire service, law enforcement and other community agencies. Parental involvement is mandatory for acceptance. Youth and parents are invited to the fire department for an initial interview to determine the youth's potential for participating in the program and the family's willingness to cooperate. At this time the program is explained and an agreement to participate is reached.	An interview is conducted with the youth and family to determine the severity of the firesetting behavior. The evaluation instruments presented in the U.S. Fire Administration's Juvenile Firesetter Handbooks may be used. All curiosity firesetters receive educational intervention. If firesetting is determined to be beyond the curiosity level, referral is made to the necessary intervention programs.	This is a highly structured educational intervention. It is divided into four age categories, ranging from 4 years old to college level. Each age category is comprised of seven fire safety projects to be completed by the youth and discussed in family safety meetings. Supervised station work time is required for older youth and a fire department visit is required for younger youth.	Youth and families classified beyond the curiosity level are referred to mental health counseling or other appropriate intervention services. Once the referral is made, it does not preclude youth and families from participating in the educational component of Partners.	After completing the seven steps of the educational component, youth are required to submit, depending on age, a project on fire safety to the fire department. A followup is conducted by the fire department within one year after completion of the fire safety project.	

Table 4.7

Examples of Juvenile Firesetter Programs (Continued)

PROGRAM LOCATION	GENERAL DESCRIPTION	COMPONENTS				
		IDENTIFICATION TO ASSIGNMENT	EVALUATION	EDUCATION	REFERRAL	EXIT TO FOLLOW-UP
Portland Fire and Rescue • 55 SW Ash St. Portland, OR 97204 • (503) 823-3615 dporth@fire.ci.port-land.or.us	The mission of this program is to identify and assess the firesetting behavior of children who have been referred to the program for the unsanctioned and/or unsupervised use of fire. It provides education intervention services and, when necessary, referral to more definitive services for the child and/or family. The program is made up of six basic elements, which include: identification, education, interview/screening, referral, follow-up, and proaction.	Children are brought to the attention of the program through fire response/fire investigation, by school officials, by mental health professionals, by child welfare professionals, by medical professionals, by juvenile court officials, or by parents/caregivers seeking assistance for firesetting behaviors in their children. The program manager begins the intake process by initiating a computer file and gathering any pertinent information from the professional identifying the behavior. The family is then contacted and an intake interview is conducted to gather information about the child, family, and incident. Educational messages are also provided and an interview appointment is usually set up. A packet of educational material, program information, and appointment confirmation is then sent to the family.	Upon meeting with the child/family, a formal interview is conducted to gather information. The "Firestoppers of Washington" interview form is used as a guideline for this process. Behavior warning signs are explored during this process. In an effort to determine and classify the motivation behind the behavior, each interview results in a needs classification. These are deemed "Little Concern," "Definite Concern," and "Extreme Concern." This assists the interventionist in determining the needs of each child/family and helps direct referral for those needing that component of the program.	Educational intervention is performed in cooperation with the Interview/Screening. Age appropriate educational strategies are used to help the child/family understand and correct the contributing factors to the firesetting behaviors. While some time is spent on fire survival skills, (Stop, Drop, and Roll; Crawl Low Under Smoke; Safe Meeting Places; etc.), most emphasis is placed on firesetting prevention issues such as Tool and Toy, Matches/Lighters in Safe Places, etc. Visual aids may be used and include video, still photos, props, etc. Each educational intervention session is customized to the child/family in question. Other adjuncts used are "Non-Fire Use Contracts" and homework assignments.	When the Interview/Screening and Education determine that the needs of the child/family are beyond what education will resolve, the program then advocates for the family by helping select an appropriate referral agency, which is already a prepared part of the program and community network. Parents/Caregivers will be asked to sign a release form to allow the program manager to communicate with the referral agency/professional. Along with predetermined referral agencies, support may also be sought from the schools, juvenile justice, and other appropriate agencies. The program remains a resource and available to continue work with the family under the direction of the referral professional.	Four months post-interview, the participating parent/caregiver who attended with the child will be contacted by phone or with a mailed survey form. The follow-up survey will question recidivism (if present, question how and other contributing factors) and customer satisfaction. Different aspects of the program, including interviewer skills will be rated by the participant. The data gathered in the follow-up phase is entered into the computer and used to perform city risk analysis to direct the Proaction component. During Proaction, the program uses the information that identifies behavioral thinking errors, geographic location, and other clues to help effectively target children with appropriate messages to prevent firesetting behaviors.

Table 4.7
Examples of Juvenile Firesetter Programs (Continued)

PROGRAM LOCATION	GENERAL DESCRIPTION	COMPONENTS				
		IDENTIFICATION TO ASSIGNMENT	EVALUATION	EDUCATION	REFERRAL	EXIT TO FOLLOW-UP
Youth Firesetter Intervention Program • City of Phoenix Fire Department-Urban Services Division 150 South 12th Street Phoenix, AZ 85034-2301 • (602) 495-5515	The mission of this program is to provide educational and counseling intervention to youth and their families experiencing problems with firesetting. There are several major components to the program. They are behavioral health, education, assessment, evaluation, Maricopa County Juvenile Court Center, Diversion Program, program awareness, and citizens advisory panel.	Youths enter the program in a variety of ways including the parent or school making a call to seek assistance, a youth who is encountered at the fire scene and referred to the program by the company officer, or a fire investigator, the police department, or any other agency making a referral. Upon receiving a request, a case worker will follow up with the family and advise them of the available resources.	When a youth and family is referred to the program, an initial intake is conducted to determine the severity of the problem. Referrals then are made to educational classes and/or to mental health. A complete evaluation is conducted by both education and mental health.	A voluntary three hour fire safety class is offered one Saturday a month. The morning group is divided into two sections: a preschool class for three to five year olds and a class for six to eight year olds. The afternoon class is for children nine years and older. Both morning classes teach fire safety behaviors. The afternoon class addresses the consequences of firesetting as well as responsibility issues incurred with firesetting. The program also offers a parenting component to discuss alternatives for parents as well as fire safety information.	Recognizing the need for behavioral health services, the Phoenix Fire Department formed a panel of mental health care providers. These services are activated on the recommendation of the intake case worker. If counseling is necessary a provider calls the family within 48 hours to set up an appointment. If insurance is not available, budgeted funds from the Fire Department will cover the cost of counseling intervention. The panel also meets on a monthly basis to discuss the case histories of firesetters and their families.	Six months after attending the fire safety class, all parents and care-givers receive a postage paid evaluation postcard. Written case summaries are completed for all youth referred to mental health.

Suggested Readings

Topic	Resources
A general reference on community-based fire education.	The National Association of State Fire Marshals. <u>The Community-Based Fire Safety Handbook.</u> Washington D.C: Rossomando & Associates.
Colorado's implementation of a juvenile firesetter prevention program utilizing the five component model of identification and assessment, evaluation, education, referral, and exit and follow-up.	Colorado Department of Public Safety (1997). <u>Colorado Juvenile Firesetter Prevention Program: Training Seminar. Volumes 1 and 2.</u> Denver, Colorado.
Fireproof Children is an intervention program featuring evaluation, education, networking, and a variety of additional resources.	National Fire Service Support Systems, Inc. (1990). <u>Fireproof Children Education Kit and Fireproof Education Handbook.</u> Pittsford, New York.
Portland Fire & Rescue is an educational intervention program that serves as a community contact point for all issues relating to fire and fire safety. They provide educational services, an interview/screening component, referral services, follow-up, and proactive educational programs to prevent firesetting behavior.	Portland Fire Bureau (1997). <u>The Portland Report '97.</u> Portland, Oregon.
Partners is a fire safety education program for juveniles and their parents using multiple teaching methods.	Palm Beach County Fire-Rescue (1997). <u>A Juvenile Firesetter Intervention Program for the Community.</u> Palm Beach, Florida.
Fire Stoppers is a youth firesetting intervention program involving a community teamwork approach with the fire service and mental health community.	Tarico, Valorie, et al. (1993). <u>Children Who Set Fires. A Manual for Mental Health and Fire.</u> Seattle, Washington: Arson Alarm Foundation.
	King County, Washington (1997). <u>Fire Stoppers of King County. 1997-1998 Program Information.</u> King County, Washington.
St. Paul's juvenile firesetter intervention and prevention program is a comprehensive approach to education and intervention.	Saint Paul Fire Department (1998). <u>Program Description and Materials For the Juvenile Firesetter Intervention and Prevention Program.</u> St. Paul, Minnesota.
The Youth Firesetter Prevention Program provides evaluation, education, and referral to counseling for firesetters and their families.	Phoenix Fire Department (1998). <u>Youth Firesetter Prevention Program.</u> Phoenix, AZ: Youth Firesetter Prevention Team, 1998.

Chapter 5

The Community Network

A juvenile firesetter program is one link in a continuum of care designed to prevent and control firesetting and arson-related activities. The continuum of care is a comprehensive network of community programs and services aimed at helping at-risk juveniles and their families. The four major points of intervention on the continuum of care are prevention, early identification, immediate treatment, and graduated sanctions. For each of these points there are corresponding community services. It is this entire network of community programs and services that help reduce juvenile involvement in firesetting and arson.

Objectives

5.1. **To describe the continuum of care as a network of community services available to juvenile firesetters and their families.**

5.2. **To present prevention programs designed to educate, support and protect youth, their families, and the community.**

5.3. **To understand the role a juvenile firesetter program plays in the early identification of at-risk juveniles.**

5.4. **To identify the community services providing immediate treatment for juvenile firesetters and their families.**

5.5. **To explain the system of graduated sanctions employed by the juvenile justice system to interrupt juvenile involvement in repeated firesetting and arson.**

Continuum of Care

The continuum of care represents a way to organize the community network of programs and services for juvenile firesetters and their families. This idea, presented in Chapter 4,

illustrates the connection between the severity of the fire behavior, the associated level of risk, the recommended intervention, and the corresponding community programs and services. As the severity of the firesetting behavior increases, so does the level of risk. Increasing levels of firesetting risk require more intensive interventions. Each type of intervention is carried out by a number of different community programs and services. The more intense the intervention, the more penetrating the community program. **Table 5.1** adds the fourth row to the continuum of care presented in Chapter 4, namely community programs and services.

For each type of juvenile fire behavior, there is a related risk level, a recognized intervention, and an identified set of community programs. For example, parents report to the juvenile firesetter program that their seven-year-old son has set the trash can in their backyard on fire for the second time this month. An evaluation classifies the youngster at definite risk, and reveals that a grandmother living with the family has recently died. The family is in a period of mourning. Fire safety education is recommended, with an emphasis on family meetings to discuss the dangers of fire to the household. Mental health intervention also is indicated, with a suggested modality of family therapy to work on the grief associated with the grandmother's death. Hence, for a youngster classified as definite risk, both education and immediate treatment is advised; and in this instance, a referral to mental health services or family therapy is necessary.

The four major points of intervention on the continuum of care are prevention, early identification, immediate treatment, and graduated sanctions. These four points represent the entire network of community programs and services. They also represent four distinct levels of community intervention. Prevention programs are designed to intercept problems before they occur. Early identification efforts are based on the premise that the sooner a problem is discovered, the better the chance of a successful remedy. Immediate treatment focuses on rapid access to the appropriate care and therapy. Graduated sanctions employ rehabilitation and correction services to change behavior. Most community programs and services working with juvenile firesetters and their families can be classified into one of these four levels of intervention.

Prevention

The basic premise of prevention is that child-set fires do not have to happen. **Table 5.2** presents three general approaches to prevention. They are education, support, and protection. Prevention efforts cover a broad spectrum of

community programs and services. All children and their parents are the target of prevention programs.

Fire safety education is the most direct method of preventing juvenile firesetting. Education programs are underway in a variety of community settings, and are sponsored by a number of national, state, and local organizations. They are designed to increase fire safety awareness through public education. Fire prevention programs can be found in schools, the fire service, and burn centers.

Table 5.1
Continuum of Care

FIRE BEHAVIOR	Fire Interest→→→	Firestarting→→→	Firesetting→→→→→→→Arson
↓ ↓ ↓ ↓ ↓	Safe→→→→→→	Unsafe→→→→→	Troubled→Delinquent→Criminal
FIRE RISK	Little→→→→→→→→→→→→→		Definite→→→→→→→Extreme
↓ ↓ ↓			
INTERVENTION	Prevention→→→→	Early→→→→→→ Identification	Immediate→→→→→→Graduated Treatment Sanctions
↓ ↓ ↓			
SERVICES	Preschool→→→→ Schools Day Care Youth Programs Fire Service Law Enforcement	Juvenile→→→→→ Firesetter Programs	Mental →→→→→→→→Juvenile Health Justice →→→→Social Services→→→→

The basic premise of prevention is that child-set fires do not have to happen!

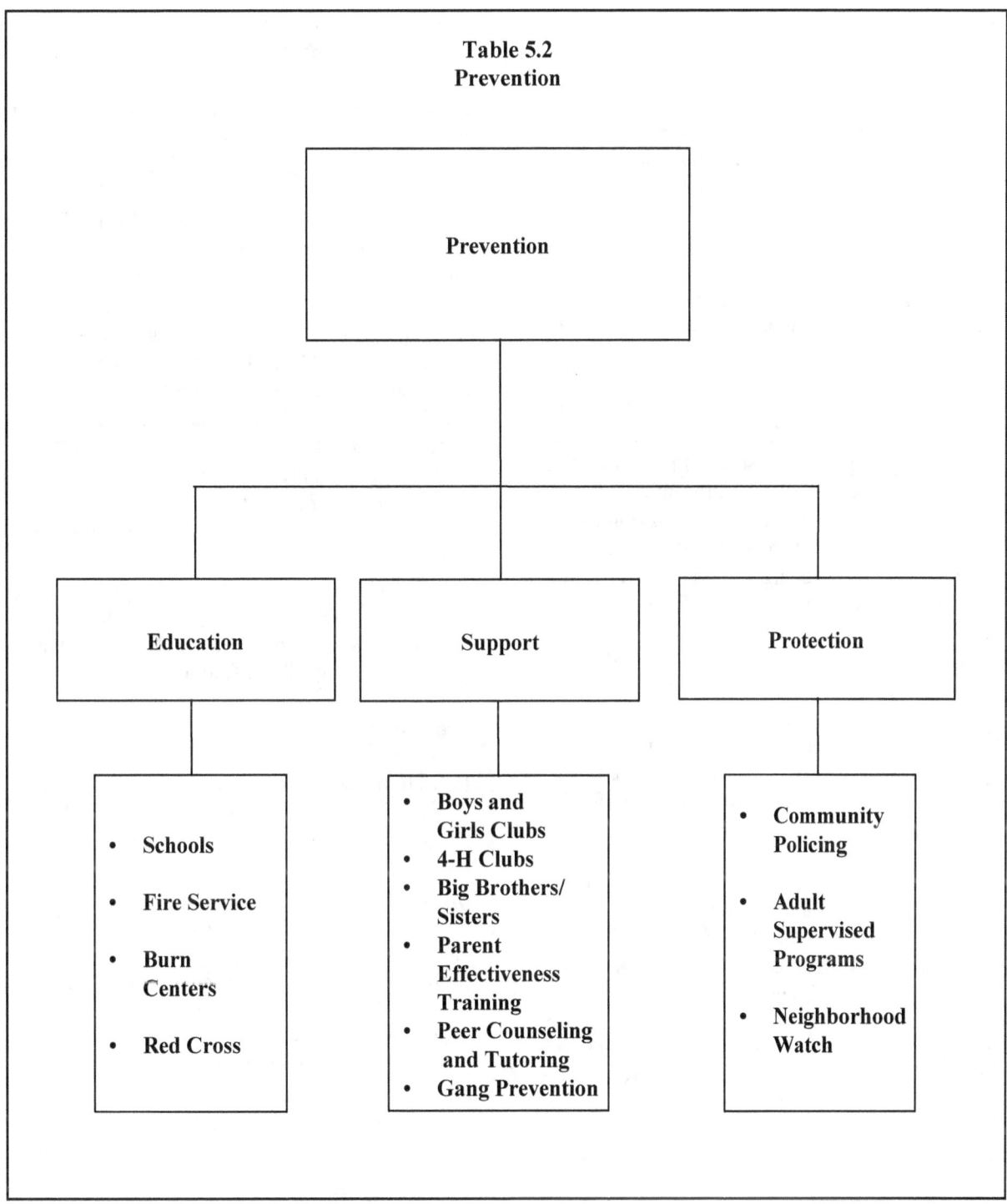

Table 5.2
Prevention

Prevention

Education

Support

Protection

- **Schools**

- **Fire Service**

- **Burn Centers**

- **Red Cross**

- **Boys and Girls Clubs**
- **4-H Clubs**
- **Big Brothers/ Sisters**
- **Parent Effectiveness Training**
- **Peer Counseling and Tutoring**
- **Gang Prevention**

- **Community Policing**

- **Adult Supervised Programs**

- **Neighborhood Watch**

Education

The most effective way to reach the greatest number of children is to offer fire prevention education in the schools. Young children (less than five years old) are at greatest risk for becoming victims of child-set fires. Therefore, pre-school fire safety education is essential. In addition, a continued fire education program through elementary, secondary, and high school can teach necessary prevention and survival skills. There are a number of excellent fire prevention programs for school settings. **Table 5.3** presents some outstanding fire prevention programs currently available for pre-school through high school.

The fire service has a long history of providing the community with effective public education programs aimed at fire prevention and safety. Fire departments participate in a variety of activities designed to raise public awareness regarding children and fire. These activities include national events and media campaigns. **Table 5.4** shows examples of fire service efforts that develop and promote prevention and safety programs in the community.

Many burn centers, providing life-saving services to victims of fire, also operate aggressive fire safety programs. Some of their activities include promoting Burn Prevention Week, mounting media campaigns during high risk times of the year, such as fourth of July and Christmas, developing parent education brochures, videos and other teaching resources, and collaborating on fire prevention programs with schools and fire departments. In addition, the Children's Hospital Burn Center in Denver operates routine educational treatment groups for at-risk firesetters and their families. This type of prevention work is critical to reducing the number of youth suffering burn injuries by child-set fires.

The American Red Cross, active in helping to provide food, clothing, and shelter to fire victims, also has recently developed its first juvenile firesetting prevention program. The Disaster Services Preparedness Bureau of the American Red Cross, in collaboration with the St. Paul Fire and Safety Services, released "Look Hot? Stay Cool!" It consists of two sections, a youth unit designed for children ages 10 to 12, and an adult unit designed for parents and caregivers of children ages 10 to 12. There are key fire safety messages taught by a collaboration of American Red Cross personnel, classroom teachers, and fire department personnel. Local Red Cross Chapters distribute the program.

Support

Support programs that improve the quality of life for children, families, and peer groups can help prevent involvement in firesetting and other delinquent behaviors. Organized youth programs that encourage children to become involved in meaningful and productive activities include boys and girls clubs, scouting groups, 4-H clubs, mentoring programs like Big Brothers and Big Sisters, church groups, the Safe and Drug Free Schools program, recreational athletic programs, such as night basketball, and other adult supervised after school programs. Family support groups such as parent effectiveness programs and family skills training sponsored by churches, YMCA's, and community centers, can help sustain a healthy home environment. Finally, older children and adolescents can benefit greatly from peer group support efforts such as peer counseling and tutoring, and gang prevention and intervention programs. Many of these activities are designed to keep youth off the streets by teaching them to educate and support other youth that may be in trouble and need help. Youth helping youth is one of the most effective methods of supporting a mutually productive and healthy lifestyle.

Protection

Education and support programs are enhanced by community protection efforts. These programs operate in conjunction with law enforcement and are designed to prevent firesetting and other criminal activity before they occur. The most pervasive example of this approach is community policing, programs designed to put more police on the streets to work with youth in schools and other neighborhood settings. These protection programs also employ educational methods aimed at specific problems, such as substance abuse prevention through D.A.R.E., or opportunities for participating in organized activities, such as the police athletic league. In addition, there are crime prevention programs organized by citizens called Neighborhood Watch. These programs detect and deter criminal activity through organized and routine citizen patrols of the community. Protecting the neighborhood and preventing criminal activity becomes a cooperative effort between law enforcement and the citizens of the community.

Early Identification

The intention of early identification is to recognize at-risk youth and prevent their further involvement in firesetting. A juvenile firesetter program is an early identification intervention. It recognizes at-risk youth, assesses their firesetting risk, and recommends further intervention. Early identification is the best chance for at-risk youth to receive effective help.

As an early identification intervention, a juvenile firesetter program serves as the major access

	Table 5.3 Prevention School Programs		
Level	**Program**	**Description**	**Source**
Preschool	Kid Safe Program	A comprehensive curriculum teaching nine critical fire safety lessons using a variety of teaching methods designed for preschoolers.	Oklahoma City Fire Department Public Education 820 NW 5th Street, Oklahoma City, OK 73106 405-297-3314
	Learn Not to Burn English/Spanish	A program guide for teachers and three resource books to help teach key fire safety and survival skills to preschoolers.	National Fire Protection Association 1 Batterymarch Park Quincy, MA 02169 617-770-3000 www.nfpa.org
	Safer Kids! A Community Action Guide For Children's Fire Safety Program	A program that includes lesson plans, games, and a video tape for preschoolers.	National Fire Service Support Systems One Grove St. #210 Pittsford, NY 14534 716-264-0840
Elementary School	Fireproof Children Education Kit	Seventy ready-to-use activities for fire safety educators and classroom teachers for students in K-6.	National Fire Service Support Systems One Grove St. #210 Pittsford, NY 14534 716-264-0840
	Learn Not to Burn	A classroom curriculum that teaches 25 key fire safety behaviors to K through eighth graders.	National Fire Protection Association 1 Batterymarch Park Quincy, MA 02169 617-770-3000
Middle School	Skills Curriculum For Interviewing with Firesetters	A 14 lesson guide for 13 to 17 year olds that identifies the causes of firesetting.	Eric Elliot 3150 Wayside Loop Eugene, OR 97477 541-682-4742
	The Science of Sizzle	A middle school science curriculum covering six areas: combustion, electricity and fire, natural gas, flammable liquids, fire in the environment, and the science of fighting fires.	F.I.R.E. Solutions, Inc. PO Box 2888 Fall River, MA 02722 508-836-9149 www.firesolutions.com
High School	Challenge for Life	A comprehensive high school curriculum that teaches fire safety and survival skills.	Georgia Firefighters Burn Foundation www.gfbg.org/challenge for life

Table 5.4
Prevention
Fire Service Programs

Method	Program	Description	Source
Community Activities	National Fire Prevention Week	A nationally coordinated effort the first week in October designed to raise public awareness about fire safety.	United States Fire Administration
	National Arson Awareness Week	A relatively new national public awareness program during the first week in May focused on arson prevention and control.	International Association of Arson Investigators
Media Campaigns	Curious Kids Set Fires	Press packet promoting national media campaign on fireplay and firesetting.	United States Fire Administration
	Big Fires Start Small	National media kit designed to explain the problem of children playing with matches.	National Fire Protection Assoc.

point to the entire network of community services. **Table 5.5** illustrates the relationship between the juvenile firesetter program and the community network. The juvenile firesetter program occupies the central position between referral sources and target agencies. The community network refers youth into the system. The juvenile firesetter program provides assessment, evaluation, and education. For those juveniles and families requiring referral for additional intervention, the target agencies are the part of the network that provide immediate treatment and sanctions.

Each juvenile firesetter program will develop their own specific linkages to referral sources and target agencies. **Table 5.5** presents a general list from which juvenile firesetter programs can select specific agencies to form their community network. Once the particular agencies have been selected, the juvenile firesetter program must establish and maintain an effective working relationship with them.

There are important steps to take to secure the linkages between referral sources, the juvenile firesetter program, and target agencies. First, referral sources and target agencies must be educated about the services of a juvenile firesetter program. This is the responsibility of the juvenile firesetter program. Referral sources must understand what types of youth they can refer and target agencies must understand what type of problems they can expect to receive.

Second, key people in the referral sources and target agencies must be identified. These key people represent not only those who will be working directly with the juvenile firesetter program, but also those who will approve the working agreement between the agencies and the juvenile firesetter program.

Third, documentation is recommended which specifies the relationship between specific referral sources, target agencies, and the juvenile firesetter program. A written agreement specifying the conditions of the service relationship and the responsibilities of each party should be authored, signed, and filed as part of the documentation of participating agencies.

Finally, there should be some mechanism to monitor the quality of services provided by the target agencies. Public or government agencies generally have to meet a certain standard of care, but private agencies are not regulated in the same manner. In addition, if private practitioners, such as mental health professionals, are offering services, they must be professionally licensed to practice in their state. Also, mental health professionals experienced in treating children and families, especially those with antisocial or delinquent behavior problems such as firesetting, should be identified to work with juvenile firesetters and their families.

It is important to have confidence in the target agencies providing recommended intervention services to at-risk youth and their families.

Immediate Treatment

Access to immediate treatment ensures that juveniles classified as definite risk receive swift and effective help. There are two major types of immediate treatment formats - mental health and social services. Referral to one or both of these formats depends on the assessed needs of the juvenile firesetter and family. A juvenile firesetter program should have strong linkages with each of these immediate treatment formats.

Mental Health

Most juvenile firesetter programs refer definite risk juveniles and their families to mental heath services. There are a few juvenile firesetter programs, such as King County, Washington and Phoenix, Arizona, that include mental health intervention as part of their program structure. However, most juvenile firesetter programs refer their cases to metal health agencies in their community.

A juvenile firesetter program selects the mental health agencies to use as their referral sources. The selection depends on factors such as resources and availability. One consideration in the selection of mental health services is whether they are public or private programs.

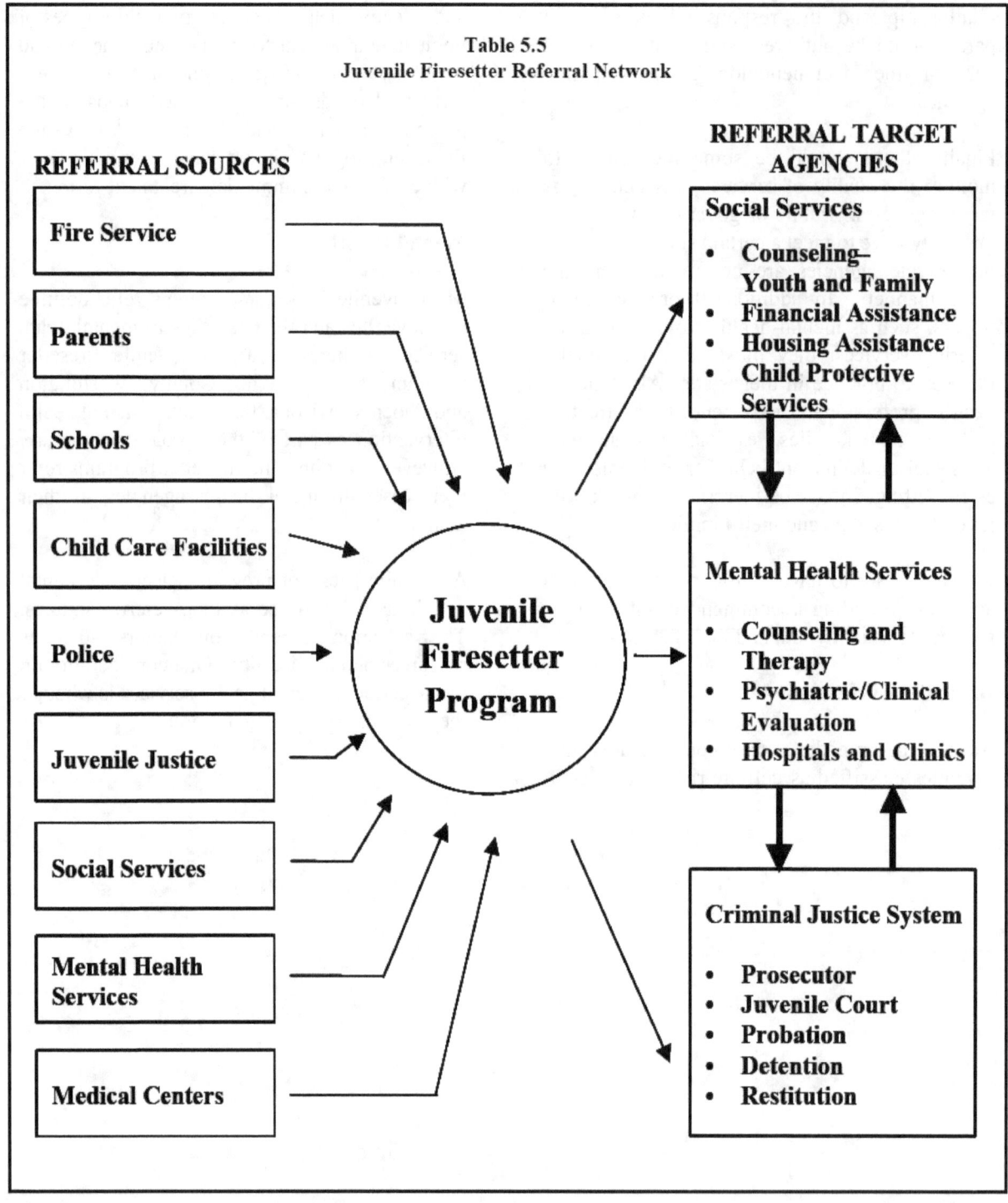

Table 5.5
Juvenile Firesetter Referral Network

Public programs operate with some state and federal support, and often provide services on an ability-to-pay basis. Private mental health programs and licensed private mental health professionals typically offer services for fees and insurance. Both public and private options are available in most communities.

A second consideration in the selection of mental health services is the type of treatment. In some communities there will be mental health professionals experienced in the treatment of juvenile firesetting. In those communities where this is not the case, it is important to refer to mental health programs and professionals specializing in the treatment of childhood and adolescent disorders. In addition, mental health interventions specializing in family therapy are effective in the treatment of juvenile firesetting.

There are a variety of successful mental health interventions for juvenile firesetting. **Table 5.6** summarizes the various treatments ranging from outpatient therapy to inpatient treatment. Juvenile firesetter programs will refer most of their definite risk cases to outpatient treatment. **Table 5.6** also lists several outpatient methods developed specifically to treat juvenile firesetting. When juvenile firesetter programs select their mental health referral sources, they should ask about the availability of these treatments.

Social Services

There are certain circumstances, namely families who are unable to support themselves, that will require a juvenile firesetter program to refer definite risk cases to social services. Federal, state, and local governments operate social services, also known as human resources, or public assistance programs. These programs provide basic services for families in need of financial assistance, counseling, or housing.

In addition, most states operate child protective services, and agency set-up to respond to reported cases of child neglect and abuse. If a juvenile firesetter program identifies a case of abuse or neglect, most states, by law, require them to report the case within 24 or 48 hours to

child protective services. This agency then investigates the case, and may or may not take action, depending on their findings. Social services programs, using public funds, provide basic assistance to families who cannot provide for themselves.

Graduated Sanctions

Graduated sanctions are designed to interrupt the progression of delinquent and criminal activity. A decision to file charges marks a youth's entry into the juvenile justice system. An effective juvenile justice system combines accountability and sanctions with increasingly intensive treatment and rehabilitation services.

Table 5.7 defines the two graduated sanctions components-rehabilitation and corrections. Each of the two components consists of sub-levels or gradations, that together provide an integrated approach. Rehabilitation is comprised of immediate therapy and intermediate sanctions. Corrections include community confinement, training schools, and aftercare.

For rehabilitation efforts to be most effective, they must be swift, certain, and consistent, and incorporate increasing sanctions, including the loss of freedom. As the severity of sanctions increases, so does the intensity of rehabilitation. At each level, if a youth continues firesetting or other delinquent activities, he/she will be subject to more severe sanctions and ultimately could be confined to a secure setting, ranging from a community-based juvenile facility to a training school, camp, or ranch.

Diversion is the main vehicle for delivering rehabilitation. A firesetting youth who is a first-time offender is likely to be placed in diversion. Rehabilitation objectives for the first-time offender include accountability (the requirement to make amends to the victim and community for the harm caused), exiting the juvenile justice system a more productive and responsible citizen, and ensuring public safety. A firesetting youth placed in diversion may be mandated to participate in community-based programs. These programs are small and open, located near the youth's home, and encourage interaction within the community. Community police officers often monitor the juvenile's progress. **Table 5.8** lists some examples of these programs.

Firesetting juveniles who are not first-time offenders or who fail to respond to rehabilitation, are subject to intermediate sanctions. Intermediate sanctions are intensive supervision programs, which are highly structured plans, consisting of short-term placement in community confinement, day treatment, outreach and tracking, routine supervision, and

discharge and followup. **Table 5.9** presents some of these program options.

An effective juvenile justice system combines accountability and sanctions with increasingly intensive treatment and rehabilitation services.

Table 5.6
Immediate Treatment
Mental Health Programs

Method	Modality	Program
Outpatient Therapy	Cognitive-Emotive	Recognition and interruption of the urge to firestart using an interview graphing technique.
	Behavior Therapy	Firesetting is abated using various behavioral methods including punishment, reinforcement, negative practice, and fantasies.
	Family Therapy	The focus is on improving and restructuring patterns of family communication and interaction.
	Group Therapy	Groups for juveniles focus on fire safety education and consequences of firesetting. Groups for parents focus on stress management and parent effectiveness training.
Inpatient Treatment	Varied	Programs offer short-term (six weeks) inpatient evaluation and treatment using satiation and family therapy, with a focus on re-entry to the family and community.
Residential Treatment	Varied	Programs offer long-term living arrangements with a highly structured format including individual and group therapy, recreation activities, and vocational training. Often a half-way house placement precedes re-entry.

Corrections

The best chance of changing the future conduct of juveniles involved in chronic firesetting and arson is to couple secure corrections with intensive rehabilitation services. Community confinement provides secure confinement in small community-based facilities that offer individual and group counseling, educational programs, medical services, and intensive staff supervision. Proximity to the community enables direct and regular family involvement with the rehabilitation process as well as phased reentry into the community, drawing upon community resources and services.

Juveniles who constitute an on-going threat to community safety or who have failed to respond to community-based corrections, may require an extended placement in training schools, camps, ranches, or other secure facilities. These programs also offer intensive rehabilitation with a focus on education, skills development, and vocational or employment training. Essential to the reentry for these juvenile is an intensive after-care program that provides high levels of social controls and on-going rehabilitation services.

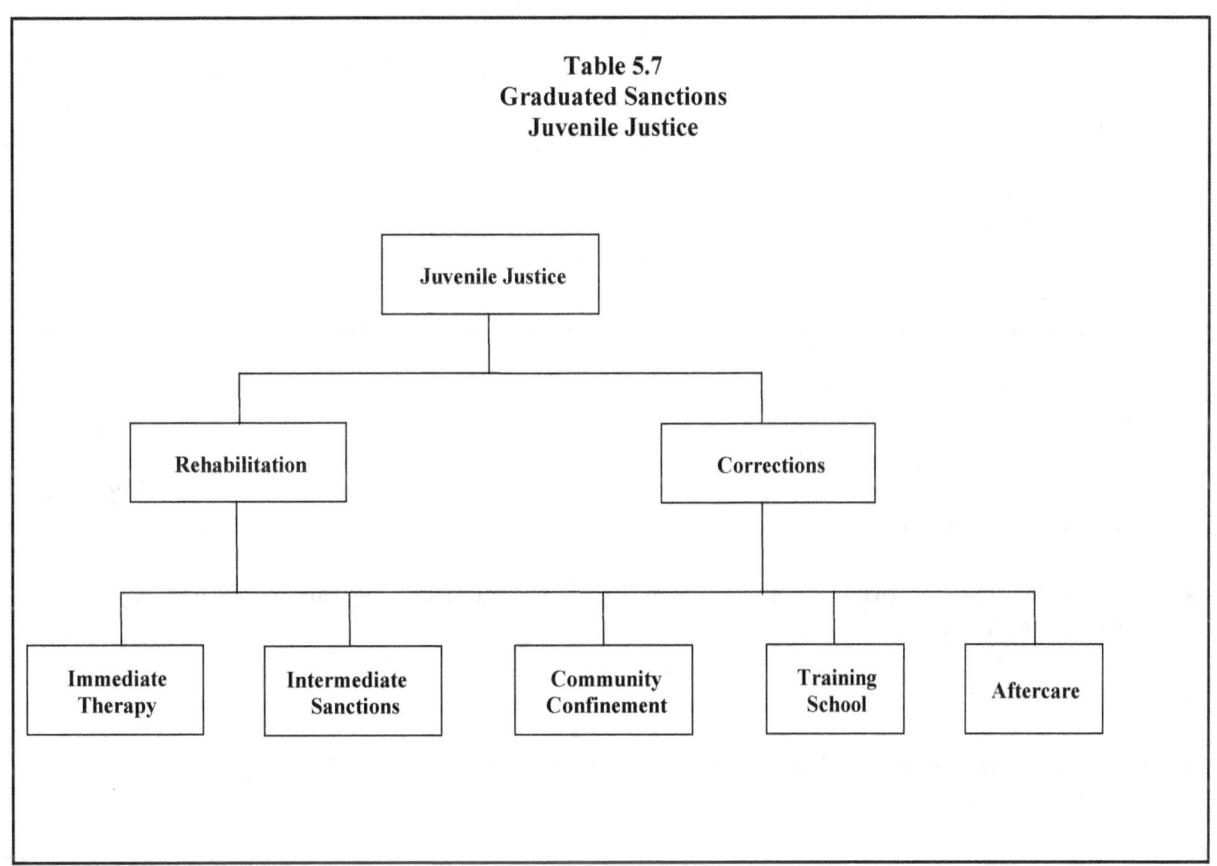

Table 5.7
Graduated Sanctions
Juvenile Justice

Table 5.8 Graduated Sanctions Rehabilitation Immediate Therapy Programs	Table 5.9 Intermediate Sanctions Intensive Supervision Programs
Neighborhood Resource Teams	Drug Testing
Informal Probation	Weekend Detention
School Counselors Serving as Probation Officers	Inpatient Alcohol and Drug Abuse Treatment
Home on Probation	Challenge Outdoor Programs
Mediation With Victims	Community-Based Residential Programs
Community Service	Electronic Monitoring
Restitution	Boot Camp Facilities and Programs
Day Treatment Programs	
Outpatient Alcohol and Drug Abuse Treatment	
Peer Juries	

Summary Points

- The most effective approach to combating juvenile firesetting and arson is an organized network of community services.

- A juvenile firesetter program can take the lead in organizing and accessing the community network of services.

- Fire safety education is essential for everyone.

- Improving the quality of family life and protecting the community are part of preventing juvenile involvement in firesetting.

Suggested Readings

Topics	Resource
A resource guide for fire departments and mental health professionals.	Burn Awareness Coalition. <u>Burn Awareness Kit</u>. Encino, CA: 1998.
An analysis of why adolescents get into trouble and what communities can do about it.	Carnegie Council on Adolescent Development. <u>A Matter of Time, Risk, and Opportunity in the Nonschool Hours</u>. New York: Carnegie Corporation of New York, 1992.
A presentation of how communities can develop and sustain effective youth programs.	Hawkins, D. and Catalano, R. <u>Communities That Care</u>. San Francisco: Jossey-Bass, Inc. 1992.
A collection of current articles on mental health programs designed to treat firesetting.	State of Colorado. <u>Juvenile Firesetter Prevention Program. Training Seminar. Volume 2</u>. 1996.
The master plan developed for the Office of Juvenile Justice and Delinquency Prevention to prevent, treat, and rehabilitate juvenile delinquents.	Wilson, J. and Howell, J.C. <u>Comprehensive Strategy for Serious, Violent, and Chronic Juvenile Offenders</u>. Washington DC: The Office of Juvenile Justice and Delinquency Prevention, 1994.

Chapter 6

Building A Juvenile Firesetter Program

A community's decision to build and sustain a juvenile firesetter program must be carefully planned before it is executed. Certain tasks must be completed to ensure that when the program opens its door for business, it offers the kinds of services necessary to help at-risk youth and their families. Building a successful juvenile firesetter program will contribute significantly to a safe and secure community.

Objectives

6.1. To present the five critical tasks necessary to start and sustain a juvenile firesetter program.

6.2. To define the specific steps for each of the five critical tasks.

6.3. To emphasize the importance of completing the first three tasks - assessment, planning, and development - before starting program operations.

6.4. To outline the procedures that will maintain a juvenile firesetter program within an effective network of community services.

Critical Tasks

To build and sustain an effective juvenile firesetter program, five critical tasks must be accomplished in a specific sequence. These tasks are summarized in **Table 6.1**. First, an assessment must be conducted to determine whether the community has a significant juvenile firesetting problem to warrant the time, attention, and resources it will take to develop a reasonable solution. Second, if the need is considered significant, then a plan must be developed for the community to take action. This planning phase results in designing the blueprint for the community's juvenile firesetter program, and involves all potential partners in

the community. The third task is program development, or setting in motion the activities that will open the program's doors for business. Once the program is implemented, the installation of a management system is essential to the effective day-to-day operations of the program. Finally, there are several activities that are recommended to sustain and enhance the longevity and success of the program.

Table 6.1
Building A Juvenile Firesetter Program
Five Critical Tasks

Task	Description
I. Community Assessment	A data-based study to determine whether a juvenile firesetting problem exists within a community.
II. Planning	The design and development of the blueprint for the juvenile firesetter program.
III. Development	Setting into place all of the components of the juvenile firesetter program.
IV. Implementation	The juvenile firesetter program opens its doors for business.
V. Maintenance	The installation of a management system to monitor the short and long-range performance of the program.

Task I. - Community Assessment

Sometimes juvenile firesetter programs originate out of the concerns of one inspired individual. These individuals may be members of the fire service who have a genuine interest in children or who have seen, first hand, the pain and damage caused by juvenile firesetting. Other times, a significant, and often tragic, child-set fire will galvanize one or more citizens to organize a community action group. However the initial interest is generated, one of the first tasks for the concerned individual or group is to acquire a detailed understanding of the juvenile firesetter problem in their particular jurisdiction. **Table 6.2** presents a summary of the steps necessary to conduct a community assessment of the juvenile firesetting problem. An accurate description of the nature and extent of juvenile firesetting and a community consensus to take action is the cornerstone of building an effective juvenile firesetter intervention program.

Step A. - Problem Identification and
Step B. - Information Acquisition

One way of determining the nature and extent of a community's juvenile firesetting problem is to conduct a needs assessment study. A needs assessment study evaluates whether there is a significant problem in a community that demands attention and action. The first step in a needs assessment study is to identify local sources that may have information relevant to juvenile firesetting. The fire department is one of the best places to start gathering information. Fire incident reports, arson investigation reports, and other statistical records can provide data on the involvement of juveniles in reported fire situations. Information on property loss, injuries, and deaths related to child-set fires can give added meaning to the numbers. Fire departments and law enforcement can provide local and national information on the number of juveniles arrested for arson. Burn centers and the Red Cross can add information related to the victims of child-set fires.

Once data on the juvenile firesetting problem is collected and organized, the interested individual or group can then evaluate the

information. This evaluation can compare data over a time span to see if there are any trends. For example, the annual percentage of child-set fires occurring in the community can be compared for a five-year period to document low and high rates. Or, the percentage of local juvenile fire incidents can be compared to state and national averages to evaluate the severity of the community problem. Local arson arrest records also can be compared to state and national rates. Given the collection of relevant data, the next step is to develop a brief needs assessment report summarizing information on the problem of juvenile firesetting.

Step C. - Consensus Building

The assessment of whether the problem of juvenile firesetting is a serious threat to the community and therefore necessitates some type of action, is likely to rely not only on the collection of data, but also on the decision-making of those interested and capable of building a juvenile firesetter program. Once a needs assessment report is prepared, it can be the focus of discussions with concerned individuals or groups. These discussions are likely to take place with the fire chief, law enforcement and juvenile justice officials, mental health professionals, community leaders, local city councils, and others to determine if the magnitude of the juvenile firesetter problem warrants community action. Discussions in these meetings are likely to focus on several topics including the cost of the problem versus the cost of the solution; whether fires set by juveniles are a significant proportion of fires set in the community, and whether juvenile firesetters are over-represented given the proportion of juveniles in the community. Reaching agreement as to whether juvenile firesetting is a significant community problem may take many meetings and discussions with a variety of decision-makers.

Table 6.2
Community Assessment

Activity	Description
Problem Identification	Developing a plan to determine whether the community has a significant juvenile firesetting problem.
Information Acquisition	Collecting, organizing, and evaluating fire incidence data to determine the extent of the community's juvenile firesetter problem.
Consensus Building	Harnessing the support of key decision-makers to reach an agreement regarding the need for a juvenile fire-setting program.

Task II. - Program Planning

If a consensus is reached to address juvenile firesetting in the community, then the next task is to develop an organized approach or plan to resolve the problem. Because each community is unique and has their own set of problems and resources, only members of a particular community can decide what constitutes a serious problem and which strategies will be most effective to address the problem. Some communities may decide that the problem is severe enough to establish a comprehensive juvenile firesetter program. Other communities may decide that the problem is only moderate, or that a separate juvenile firesetter program is beyond the resources of their community. In these cases, the decision may be made to bolster existing programs or to add one or two new features to their current activities. Whatever the decision, there are a number of planning steps that can be taken to ensure an effective outcome for the community. **Table 6.3** summarizes the five steps that comprise program planning.

Table 6.3
Program Planning

Activity	Description
Problem Definition	An agreement that the extent of the community's juvenile firesetting problem warrants building a juvenile firesetter program.
Leadership	The selection of someone responsible for running the program.
Location	The identification of the primary location to house the juvenile firesetter program.
Site	The determination of the program's geographic boundaries and jurisdiction of service delivery.
Resources	The cost estimate to operate the program and the identification of potential funding sources.

Step A. - Problem Definition and
Step B. - Program Leadership

If the procedures for community assessment have been followed, then the first two steps of program planning, problem definition and program leadership, are well underway. A needs assessment study summarizing the nature and extent of juvenile firesetting in the community defines the problem. Reaching a consensus amongst community decisionmakers to take action identifies the key players and perhaps the program leadership. However, it is important at this stage to assign the primary leadership responsibilities to one individual. Typically this person will be associated with the fire service, although law enforcement, juvenile justice and mental health professionals also can assume this role. Naming a leader marks the beginning of the juvenile firesetter program.

Step C. - Program Location

The selection of the program leader generally determines the primary location of the juvenile firesetter program. The majority of juvenile firesetter programs are administered by and located in the fire service. Juvenile firesetter programs can be placed in many different branches of the fire service, including public education, fire prevention, arson investigation, the Office of the Fire Chief, and the Office of the Fire Marshal. The primary reason for housing juvenile firesetter programs within the fire service is that they are in the best position to identify juvenile firesetters. The local fire department is the agency most likely to be called by parents recognizing firesetting behavior in their children; they are the first to arrive at a fire scene, and most likely the first to recognize juvenile involvement in the fire. They are also well-connected to other community agencies, such as the schools, that are likely to refer firesetting youth for help. In short, the fire service represents the best frontline for identifying at-risk youth.

If fire departments do not have the resources to house a juvenile firesetter program, but there is a recognized community need, then there are other options for locations. Some communities have recruited private agencies or non-profit organizations to establish their juvenile firesetter programs. Also, mental health and counseling agencies have established juvenile firesetter programs. Finally, there are some juvenile justice agencies that operate juvenile firesetter programs as part of their diversion effort. Even if the juvenile firesetter program is located outside the fire service, they are still likely to work closely with the fire department to ensure that all at-risk youth are identified, evaluated, and have access to appropriate intervention.

Step D. - Program Site

The next step in program planning is to specify the geographic area that will be serviced by the juvenile firesetter program. Most juvenile firesetter programs operate within their fire service district, that is, their local community. However, there are other programs that operate over a larger geographic area, including one or more counties as well as entire states. One of the advantages of working over a larger geographic area is that many referral agencies, such as mental health, social services, and juvenile justice, operate at the countywide or state level. A juvenile firesetter program that spans one or more counties can combine resources instead of competing for them. There are many states, Colorado, Massachusetts, and Oregon, to name a few, that are organized both at the local as well as the statewide level. This coordinated network of programs and services ensures that all at-risk youth within a given state will have an excellent chance of receiving the necessary type of intervention.

Step E. - Resources

Few programs can operate without funding. In the planning stage it is necessary to identify sources willing to commit monies to the program. The closely related topic of staffing, usually the program's greatest expenditure, also must be addressed. The number of paid versus volunteer positions directly impacts the budget, and in part, determines whether the juvenile firesetter program will run on a shoe string or have the resources to operate on a more substantial and stable level.

Task III. - Program Development

The program development task puts into place the various components of a juvenile firesetter program. Once the components are in-place, the program is ready to begin its work in the community. **Table 6.4** presents the steps that will result in the development of an effective juvenile firesetter program.

Step A. - Management Team

The first step in program development is to craft a management team. Typically the juvenile firesetter program leader heads the management team. If the fire service is the site selected for the program, then this person is likely to report to someone in the fire department's chain of command. Exactly who this is will depend upon the particular unit within the fire department that houses the juvenile firesetter program. The program leader also may elect to identify one or more co-workers as assistants. Together they will form the management team responsible for running the day-to-day operations of the program. Some of these responsibilities include being in charge of the various components of the program, either directly or by supervising others who are assigned to provide assessment, evaluation, education, referral, and follow-up services. The management team also will be responsible for ensuring that the juvenile firesetter program is operating as an integral part of its fire department. Linkages must be established between other department programs such as fire prevention and arson investigation as well as with the department's chain of command. In addition, the management team must build the necessary linkages between the juvenile firesetter program and the other agencies comprising the network of community services. The management team is responsible for the mechanics of running the juvenile firesetter program as well as for the leadership and direction of the program.

Step B. - Advisory Council

To create strong linkages between the juvenile firesetter program and the network of community services, it is recommended that one of the first actions the management team should consider is establishing an advisory council. The primary role of an advisory council is to facilitate multi-agency cooperation in planning, implementing, and maintaining the community's juvenile firesetter program. The council should be composed of representatives from all agencies in the community whose responsibilities relate to juvenile firesetters. The council can include representatives from the fire service, law enforcement, firefighter unions, mental health, burn centers, social services, the schools, juvenile justice, and the media. The importance of working with the media will be covered in a section to follow on public awareness.

The juvenile firesetter program leader typically takes the responsibility for recruiting the members of the advisory council. Because the program leader is often a member of the fire service, they can represent the local fire department on the council. The program leader should contact the administrator of each key community agency to explain the juvenile firesetter program and the need for developing the advisory council. Meetings should be set-up, if possible, between the juvenile firesetter program manager and potential council members. The needs assessment report and other relevant written materials also should be distributed to potential members. Once the first agency administrator agrees to participate on the council, it will be easier to convince others to come on board. Once the members of the council have agreed to participate, a general meeting can be organized. During this meeting, members can develop procedures directing the future conduct of the council.

There are several important reasons for developing an advisory council. First, if the key agencies comprising the network of community services are represented on the council, then the first step has been taken in organizing a coordinated system of delivering services to juvenile firesetters and their families. This council can help ensure that at-risk youth will not fall through the cracks, but will receive the necessary and appropriate intervention services. Second, members of the council can help to identify potential funding sources to help support the operation of the community program. Third, the council can work to clarify the roles of each agency in their delivery of services. Council members can educate each other about how their agency can effectively work with juvenile firesetters and their families. For example, the fire service can be designated as responsible for providing assessment, evaluation, and education, while mental health can be responsible for providing counseling. Fourth, the council also can help develop the specific referral agreements and determine how they will operate between community agencies. Finally, council members can identify other agencies or individuals in the community that work with firesetters. They can distribute information about the juvenile firesetter program to their agencies and to other agencies within the community. The council members can serve as prominent advocates in their community for their juvenile firesetter program.

Step C. - Service Delivery System

Each juvenile firesetter program will select the type and range of services it will provide to its community. Because a juvenile firesetter program is an early identification and intervention effort for at-risk youth, building the five core service components is recommended. To review, these five components are identification to assignment, evaluation, education, referral, and exit to follow-up. They have been described in detail in previous chapters. The juvenile firesetter management team is responsible for building these program components. In addition, in collaboration with the advisory council, the juvenile firesetter management team is responsible for building the community network of services for juvenile firesetters and their families.

Juvenile firesetter programs will develop differently, given the characteristics, needs, and resources of their community. Not every community may be able to offer all program services. The core program components are likely to be structured differently from community to community. Nevertheless, in the program development phase, building the specific program components to service at-risk youth establishes the operation of the juvenile firesetter program in the community.

Step D. - Budget

To solidify the development of the juvenile firesetter program, the management team must outline the estimated costs of starting-up and running the program. A line-item budget specifying the program costs allows for careful planning of the program's impact on current operations. Developing an annual budget for the juvenile firesetter program represents responsible program planning.

Although budgets will vary depending on how each community decides to build their juvenile firesetter program, there are some common categories of costs that most programs share.

The major categories are personnel and non-personnel costs. Personnel costs reflect the salaries and associated benefits of those assigned to provide services to the juvenile firesetter program. There are many ways of assigning these costs. Frequently, personnel costs are borrowed from other programs already in existence. Personnel costs may be traded for compensation time or other forms of non-monetary compensation. Sometimes, professionals will donate part of their time to the program. Costs will depend on the level of the personnel assigned to manage and provide services to the program. For example, there is a range of individuals in the fire service that can be assigned to manage the juvenile firesetter program. This range, from firefighter to captain, will determine the actual personnel costs attached to the program. Non-personnel costs include those items and procedures necessary to sustain the day-to-day operations of the program. Office supplies, copying costs, computer expenses, and evaluation and education materials, are some of the expected expenditures. There are many ways to fund these costs, which will be the topic of the following section.

Table 6.4
Program Development

Activity	Description
Management Team	The selection of a team to support the work of the program leader.
Advisory Council	Solicit the leadership of key community agencies to form a group that can offer referral services as well as consultation to the developing juvenile firesetter program.
Service Delivery System	Specify the program components - identification, evaluation, education, referral, and follow-up that will be offered by the program.
Budget	Estimate the costs of the program's services.
Funding	Establish sources of financial support, donations, and contributions to the program.
Organizational Chart	Specify in writing the various organizational relationships between program management, the advisory council, and the service delivery system.
Interagency Linkages	Building an effective, multi-agency community network of services for juvenile firesetters and their families.

The annual budget for the operation of the juvenile firesetter program specifies the estimated costs of program operations. During the first year, there may be start-up costs that will not be included in budgets for subsequent years. For example, there may be costs attached to training service providers during the first year of program operation. These costs often are one-time expenditures, which are not incurred in following years because experienced service providers can train new staff as they join the program. An accurate estimate of the cost of running a juvenile firesetter program is critical

to convincing decision-makers of the value of the program to the community.

Step E. - Funding

Once the budget is estimated for a juvenile firesetter program, the next task is to fund the program. The operation of a juvenile firesetter program depends to a large extent on available resources. Public and private monies are the two basic resources for funding. One or both of these methods can be used to support a juvenile firesetter program.

Since most juvenile firesetter programs are run by the fire service, and public monies support the fire service's budget, some part of the funding for a juvenile firesetter program usually comes from the public monies. Public monies are those funds that support local, state, and national programs through the use of tax dollars. Public monies also support mental health programs, social services, and the juvenile justice system. In addition to monies allocated to fund programs, many state and federal agencies have special contracts and grants they award to individuals or community agencies proposing to start new programs. Therefore, it is important to consider not only the routine funding sources of the fire service, but also those of related state and national agencies that could support building a juvenile firesetter program.

Because the problem of juvenile firesetting effects the entire community, private companies, community organizations, and service groups often are willing to support juvenile firesetter programs. This support may be financial or it may come in the form of donations or in-kind contributions. Companies can donate their program planning advice, management expertise, public relations assistance, and fund-raising services. Donations and in-kind contributions can take the form of office supplies and materials, computer equipment, and printing costs. Community organizations and service groups can provide volunteer time. There are several private companies that have supported juvenile firesetter programs, including the insurance industry and companies marketing child-resistant lighters. Community organizations such as Boys and Girls Clubs and Big Brothers/Sisters, and service groups such as the Kiwanis and Shriner's, all have become involved in juvenile firesetter programs. If these organizations understand that reducing juvenile involvement in firesetting reduces property loss and saves lives, then they are likely to lend their support to making their community a safer place in which to live.

It is recommended that juvenile firesetter programs consider a strategy that combines both public and private resources. **Appendix 6.1** presents a list of several public and private organizations that support juvenile firesetter programs. A public/private partnership allows for a number of different organizations to lend a helping hand toward building a juvenile firesetter program for their community. In this way, both the public and private sector have a stake in the juvenile firesetter program and they can work together to make it a successful enterprise.

Step F. - Organizational Chart

At this point in the development of a juvenile firesetter program, it is a good idea to draw-up an organizational chart illustrating the operation of the program. Although the structure of each juvenile firesetter program will look differently on paper, there will be some common elements among programs. With consultation from the advisory council, the organizational relationships among the management team, the advisory council, the primary program site, and the service delivery system should be specified. Understanding how these various program operations are connected will clarify the working relationships they will have with one another. **Appendix 6.2** contains some current examples of juvenile firesetter program organizational charts.

Step G. - Interagency Linkages

Linkages between referral sources, the juvenile firesetter program, and referral agencies must be clearly defined, and preferably codified by a written agreement or contract. Referral sources are those parents, caregivers, and community

services, including different departments within the fire service, that direct juveniles into the program. Referral agencies are those community services, such as mental health, social services, and juvenile justice, to which the juvenile firesetter program will direct its definite and extreme risk cases. The juvenile firesetter program is responsible for securing and maintaining the linkages between referral sources, the program, and referral agencies. These linkages represent the community network designed to help juvenile firesetters and their families.

Building these linkages can be a time consuming process, but with the advisory council in-place, a major step has been taken in building this network of community services. While the council members usually are not the people responsible for the day-to-day working relationships between their organization and the juvenile firesetter program, they are the conduits for reaching the person who will help to define and establish these linkages.

Chapter 5 outlines the steps that must be taken to maintain these linkages including understanding the role each agency plays in working with juvenile firesetters and their families; identifying the key people within each agency who will be working with the youth; documenting the relationship between agencies, and monitoring the quality of services provided by the agencies. How each of these steps is administered depends on decisions made by the management team in conjunction with the advisory council.

Task IV. - Program Implementation

Now its time to put the planning and development of the juvenile firesetter program into action. Before the doors are open for business, there are implementation steps that must be completed. **Table 6.5** summarizes these five steps.

Step A. - Staffing

The size of the staff will depend directly on the size of the juvenile firesetter program. For small programs, there may only be a program manager who handles the majority of the responsibilities. In larger programs, there may be a management team that selects additional staff. In addition to the size of the program, the site of the program also will influence staffing patterns.

If the juvenile firesetter program operates within the fire service, there are a number of ways the program can be structured and staffed. Fire service staff such as fire investigators, fire educators, and firefighters can provide services. The type of fire service personnel selected to participate in the program will depend on where, within the fire service, the juvenile firesetter program is housed. For example, if the program is operated by fire investigation, then it is likely that fire investigators will be involved in identifying, assessing, and evaluating juvenile firesetters and their families. Or, after the initial identification has occurred, fire investigators may refer juvenile firesetters to fire prevention. Fire educators then can evaluate, educate, refer, and follow-up these at-risk youth. Another option is that all juveniles are seen by on-duty firefighters assigned to the juvenile firesetter program. These firefighters are responsible for providing the services to these juveniles and their families.

How staff is selected and assigned will vary not only with respect to where the program is housed within the fire service, but also with respect to available resources. In addition, the structure of the program, that is, how the service delivery system is set-up, will influence staffing patterns. There are a number of decisions, such as whether the program will be operated by full or part-time staff, whether they will be paid or volunteer, or whether it will be some combination of these factors, that will need to be addressed. These issues also need to be considered for those agencies involved in the community network. Staff selection and compensation will be important questions for referral agencies such as mental health, social services, and juvenile justice.

Table 6.5
Program Implementation

Activity	Description
Staffing	The selection of staff to provide the program's services.
Training	The development of a training program for all new management and staff.
Documentation	Setting-up a written or automated record-keeping system for all cases entering the program.
Confidentiality	Program policies must be established regarding the privacy of written and verbal communications, access to and sharing of records, and media involvement.
Liability	The program must be protected from potential legal action because of the behavior of firesetters and their families.

If the juvenile firesetter program operates outside of the fire service, the program site, size, and structure also will determine staffing patterns. Programs operated by other organizations such as non-profit groups, hospitals, mental health facilities, and juvenile justice programs, will have unique management and staffing structures. They will, however, have certain staffing patterns in common because of the juvenile firesetter program components - identification, evaluation, education, referral, and follow-up - they offer to their community.

Step B. - Training

Regardless of the professional background of the selected management and staff, they all should receive standardized training. A training workshop should occur prior to opening the juvenile firesetter program's doors for business.

It is usually the responsibility of the program manager to organize this workshop. In addition to inviting all staff, members of the advisory council and personnel from the referring and referral agencies should attend the workshop.

There are a number of decisions to make when designing the format for the training workshop. Where the workshop is held, when, and for how long are key issues. The juvenile firesetter program site is often selected for the location of the workshop. The workshop room should be large and comfortable enough to accommodate everyone. Depending on the number of participants, this type of room may not be available at the program site. Hence, another location, perhaps at one of the referral agencies or other nearby community resources, might work. The timing of the workshop will depend, in part, on the participants' availability as well as when the program wants to begin offering services. Most training workshops run for a day, although half-day and two-day workshops also have been conducted. The program manager, in collaboration with the advisory council, will make most of these decisions.

The topics presented during the training workshop should focus on helping participants understand their respective roles in providing services to the juvenile firesetter program. **Table 6.6** presents a selection of recommended topics. The program manager, in collaboration with the advisory council, should select those topics considered critical to the operation of their juvenile firesetter program.

Who will conduct the training is another important decision. National experts can provide information on most all of the topics selected by the program manager. However, many workshops also benefit from a team-teaching approach in which the national expert is assisted by personnel from local agencies such as mental health, social services, and juvenile justice. The advantage of this approach is that resident specialists can contribute their knowledge of those local issues relevant to building their community's juvenile firesetter program.

There are several excellent resources to use in a training workshop. Participants should be provided with at least one written document outlining the parameters of the juvenile firesetter program. Eventually each juvenile firesetter program should have its own program handbook. However, it is unlikely that, at the beginning of a program, such a handbook will be developed. Therefore this Handbook can be the

---

Table 6.6
Selected Training Topics
Juvenile Firesetter Program Workshops

- **National, state, and local statistics on juvenile firesetting and arson.**
- **Personality profiles of juvenile firesetters and their families.**
- **How to interview juvenile firesetters and their families.**
- **How to assess and evaluate juvenile firesetters and their families using the Risk Surveys and the Comprehensive FireRisk Evaluation.**
- **How to determine firesetting risk levels.**
- **Special case studies of juvenile firesetters.**
- **Educational programs for juvenile firesetters and their families.**
- **How to build an effective network of community services for juvenile firesetters.**
- **Developing budgets and funding proposals for juvenile firesetter programs.**
- **How to build publicity and outreach activities into your juvenile firesetter program.**
- **Management information and evaluation systems for juvenile firesetter programs.**

---

primary teaching resource for the training workshop. In addition, The United States Fire Administration (USFA) has developed three specific training packages to help communities build their juvenile firesetter programs. They are a Community Awareness Program, a Professional Development Training Program, and a Practitioner's Training Program. **Appendix 6.3** presents a detailed description of each of these training resources. Each of these packages contains multi-media presentations and can be used as train-the-trainer programs. All three packages are distributed free by the USFA and can be obtained at their web page: www.usfa.fema.gov.

Step C. - Documentation

It is the responsibility of every program to document its work with juvenile firesetters and their families. Documentation begins at the point the juvenile is identified to the program. At the point of entry, each case must be logged into the program. There are several ways this can be done. **Appendix 2.1** presents examples of juvenile firesetter initial contact forms.

For each juvenile and family there should be a case record and an assigned identification number. As each case moves through the juvenile firesetter program components, from identification to evaluation, education, referral, and follow-up, there should be documentation of all actions. At evaluation, completion of the Risk Surveys (**Appendix 3.1**) and/or the Comprehensive FireRisk Evaluation (**Appendix 3.2**) should become part of the case record. If there are other assessment or evaluation documents, such as a fire incident report, they also should be included in the case record. If education intervention occurs, a written summary should be included in the case record. If one or more referrals are made, these also should be documented. Finally, all follow-up contacts should be recorded. The case record documents the path of every juvenile and family from the time they enter to the time they exit the juvenile firesetter program.

These case records can be kept in written form in files secured in a locked drawer. They also can be automated and secured in computer files with password protection. Access to these files is limited, and will be discussed in the following section.

Step D. - Confidentiality

There are four major questions regarding confidentiality when working with firesetting juveniles and their families. Who has access to the case records of the juvenile firesetter program? How confidential is exchanged verbal communication? How do you protect shared confidences? When and how can you protect confidential communication from the media?

1. Access to Case Records

The case records of the juvenile firesetter may contain sensitive information on a variety of topics related to juveniles and their families. Only authorized program staff should have access to these files. Access by anyone else is limited and depends on certain factors. If a court of law subpoenas files, then the program must comply by turning over the records. These files then become the property of the court. If a person or agency outside the program requests the records, specific procedures must be followed before they are released. Because these are records of minors, disclosing information from their records should be discussed with their parents. In some states written parental permission must be obtained before the information is released. Because laws regarding the sharing of juvenile files vary from state to state, it is important for each juvenile firesetter program to consult with their local district attorney to clarify the conditions under which these files can be shared with other individuals and agencies. An example of a release of information form is included with the Risk Surveys in **Appendix 3.1** and the Comprehensive FireRisk Evaluation in **Appendix 3.2**. In general, the information in case files should be disclosed when there are a specific set of circumstances and reasons suggesting that disclosures will be in the best interest of the juveniles.

2. Confidentiality of Verbal Communication

The second area of concern is the confidentiality of verbal communications between the juveniles, their families, and the service providers of the program. For example, during an evaluation interview some youth may want to confide in their interviewers and tell them things they do not want their parents to know. In addition, parents may put pressure on the interviewers to tell them all about what their children have said, or, parents, themselves, may want to share information in confidence. In these circumstances, it is important to build a reasonable trust. The idea of a reasonable trust is that everyone has a right to private thoughts and feelings, and that confidential communications will not be disclosed unless it is in the best interest of the juveniles and their families. Also, the disclosure of confidences will not occur without the person's knowledge. Therefore, if information is shared in confidence, the trust will be kept, unless there is an important reason, such as a threat to harm or destroy person or property, to break the trust. Before a trust is broken, the juveniles or family members whose confidence is being broken should be informed and the reasons why clearly stated. Complete confidentiality in any relationship is somewhat unrealistic, but building a reasonable trust lays the groundwork for an effective working relationship.

3. Protecting the Confidence

Once a reasonable trust is established, the third area of concern is protecting the confidence. Firesetting is often an embarrassing and painful event in the lives of juveniles and their families. Most citizens living in the community, including relatives and close friends, do not easily understand or accept firesetting. Therefore, there are circumstances in which juveniles and their families may want their privacy protected. For example, during an evaluation interview a decision may be made to contact the juvenile's school for information. Before this occurs, it is necessary to speak with the parents about this and obtain written permission regarding the release of information. Issues of whether school authorities know about the firesetting and whether they "should" or "need" to know must be discussed with parents. Although open communication channels are helpful to everyone, there is the potential risk that disclosure of certain types of information, such as a history of firesetting, may negatively label juveniles and deny them future learning or work opportunities. Therefore, in each instance where there is a potential to disclose information, parents need to be informed, and issues of privacy and confidentiality must be carefully considered on behalf of the juveniles and their families.

4. Confidentiality and the Media

The final area of concern is the disclosure of the identity of firesetting juveniles and their families to the media. Because juvenile firesetting sometimes results in damaging and costly fires within the community, the media becomes interested and concerned with describing the stories of the fires. A juvenile firesetter program is likely to receive requests from the print and television media for interviews with these juveniles and their families. The issues of whether to grant interviews and reveal their identities are two different decisions, both of which rest with the juveniles and their families. However, it is the responsibility of the juvenile firesetter program to inform them of the risks and benefits associated with granting interviews and revealing their identities. The major risk of granting interviews and revealing identities is that the public identification of these juveniles and their families could result in negative reactions from family, friends and associates, and place a stigma on them within their community. The potential benefit is that other juveniles and families suffering from the same problem will come forth, and as a result of this disclosure, seek the necessary help to prevent another fire tragedy. If, after careful discussion and consideration, juveniles and families decide not to grant interviews, the program cannot release any case material or information. If the decision is made to grant interviews, but not to reveal identities, then the program should facilitate the interviews. There should be a written agreement between the juveniles, their families, the program, and the media as to exactly how the identities of those involved will be protected. All written case material released by the program should not have any identifying markers. Finally, if juveniles and families agree to interviews revealing their identities, then written parental permission releasing this information must be secured. The juvenile firesetter program must guide juveniles and families in weighing the risks versus the benefits in working with the media.

Step E. - Liability

Liability refers to the potential for juvenile firesetter programs to be at risk for legal action because of the behavior of firesetters and their families. It is important for programs to protect themselves from being held liable for the action of juvenile firesetters. There are two steps that can be taken to handle this problem.

First, liability waivers that release programs from being responsible for the actions of juveniles can be developed and implemented. Juvenile firesetter programs should develop these written waivers in consultation with their local district attorney to make sure that all the pertinent areas of concern are addressed. The parents of juveniles entering the program must sign these waivers.

Secondly, juvenile firesetter programs should know whether they have insurance to cover the risks that can arise when working with juvenile firesetters. For example, what if a youth sets fire to part of a building in which they are receiving services? Programs need to ask if there is insurance coverage for this type of event, and if so, is it adequate? Substantial insurance coverage is another way to protect juvenile firesetter programs and their ability to provide services to at-risk youth and their families.

Task V. - Program Maintenance

Once the doors have been opened for business, there are certain activities that will help sustain and strengthen the juvenile firesetter program within the community. These activities are presented in **Table 6.7**. It is the responsibility of the juvenile firesetter program to be concerned about the short-term survival, that is running the day-to-day operation of the program, as well as providing a high quality of continued service to the community.

Step A. - Operations Handbook

A juvenile firesetter program must document their day-to-day operations. The purpose of an operations handbook is to develop written documentation of program procedures. Most

juvenile firesetter programs design their own handbooks. The organization of these handbooks varies from program to program, but most all of them describe specific procedures for identification, evaluation, education, referral, and follow-up. The program manager usually is responsible for the development of the handbook in collaboration with program staff. The advisory council is consulted during its development, and often approves the handbook prior to distribution. All program leaders, management, supervisors, staff, and members of the advisory council should receive copies of the handbook. If the program is operated within the fire service, the Fire Chief, captains, and any other individuals in the chain of command also should have copies. In addition, the handbook can be used as the primary training resource for new personnel as they join the program.

Step B. - Resource Directory

Many juvenile firesetter programs have developed a resource directory for their community. A juvenile firesetter directory contains the names, addresses, phone numbers, and e-mail addresses of agencies that work with juvenile firesetters and their families. The directory can include local, county, and state-wide agencies. In the case of long-term inpatient or residential treatment facilities, because there are so few who work with juvenile firesetters, it may be necessary to list national resources. Members of the advisory council should be able to provide much of the information needed for the directory. Additional resources can be obtained by communicating with local or countywide fire departments, mental health agencies, and social services asking for their help in identifying resources. The resource directory is most useful to the juvenile firesetter program when referring youth and their families for services outside the program.

**Step C. - Monitoring and
Step D. - Evaluation**

Once the juvenile firesetter program is serving the community, it is important for program management to be able to monitor the level and volume of juveniles and families that come through the doors. Having current and accurate data such as the number of cases handled, case type, firesetter characteristics, and number and type of services rendered, provide management with information on program operations. This information can be used to determine the relative impact and effectiveness of the program.

Before a juvenile firesetter program builds their information system, program leadership must ask questions regarding the application of the information. For example, will the information be used to convince funding sources to sustain or increase the program's budget? Or, will the information be used to describe the types of at-risk youth and families receiving services? There are several ways to collect and analyze information; therefore careful attention should be paid to developing a reporting system that will strengthen program operations.

There are two types of information systems. A Management Information System (MIS) and an Evaluation System (ES). A MIS will summarize the program's caseload, track, and report the number and type of program activities, and provide data for annual reports, funding agencies, and evaluations. **Table 6.8** presents the categories of data collected for a MIS. Extending the MIS to include recidivism and other follow-up data provides the building blocks for an evaluation system (ES). An ES is an extension of the MIS, and contains all of the MIS information plus follow-up data on firesetting recidivism and other events such as school or family problems, arrests, etc. An evaluation system also can monitor information on the average dollar loss per year resulting from juvenile-set fires. With this information comparisons and trends can be tracked to assess the impact of a juveniles firesetter program on reducing the costs of juvenile-set fires. **Table 6.9** outlines the specific data collected in an ES. While a MIS summarizes information on program operations, an ES provides data on the impact of the program on reducing the incidence of juvenile firesetting.

Table 6.7 Program Maintenance	
Activity	**Description**
Operations Handbook	This handbook documents the day-to-day operations of the juvenile firesetter program.
Resource Directory	This directory lists the names, street addresses, phone numbers, and e-mail addresses of all agencies that work with juvenile firesetters and their families in the geographic area serviced by the juvenile firesetter program.
Monitoring and Evaluation Systems	Building these systems into a juvenile firesetter program allows management to monitor the level and volume of the caseload as well as the quality of received services.
Public Awareness	A publicity and outreach effort educates the general public as well as specific target populations about the problem of juvenile firesetting and informs them about the availability of the program.
Continuing Education	Continuing education opportunities for management and staff will ensure that the program operates with the most current knowledge and information and maintains a high standard of performance.

Table 6.8 Management Information System Data Collection	
Data Category	**Information**
Case Characteristics	-Referral Source -Age, Sex, Race -Family Status -Details of Current Fire Incident -Details of All Past Fire Incidents -Risk Level
Services	-Educational Services -Referrals -Mental Health Services -Social Services -Juvenile Justice -Other Referrals
Case Disposition	-Outcomes of Services -Juvenile Justice Status
Program Activities	-Education and Prevention -Training -Resource Material -Development -Media

Table 6.9 Evaluation System Data Collection	
Data Category	**Information**
Firesetting Recidivism	-Resource of Firesetting
Delinquency	-Arrests -Probation -Conviction -Incarceration
School	-Academic Problems -Disciplinary Problems -Truancy -Expulsion
Mental Health	-Contacts
Social Services	-Contacts
Family Environment	-Discipline
Personal	-Functioning
Costs	-Average Dollar Loss Per Juvenile Firesetter

Whatever type of information system a juvenile firesetter program builds, they will have to resolve certain key issues. Some of these concerns include who will design the system, will it be manual or automated, where will it be housed, and who will be responsible for running it? These important questions will determine the usefulness of the information system to the juvenile firesetter program.

Step E. - Public Awareness

It is the responsibility of the juvenile firesetter program to inform the community of the availability of its services. There are two major activities that will increase the visibility of the program within the community. A publicity program can take advantage of exposure through a number of different types of media. A community outreach effort can ensure that specific target populations in need of services will be aware of the program. If a juvenile firesetter program increases public awareness through publicity and community outreach, then it must anticipate and accommodate an increase in the number of at-risk youth and families requesting its services.

An effective publicity program will increase the public's general awareness about the problem of juvenile firesetting and the solutions available in the community. A publicity program can have several objectives, ranging from educating the general public about the seriousness of the problem to announcing the availability of services. The three most commonly used communication modalities are print media, television and radio, and press conferences. Each of these modalities employs various types of strategies and has associated outcomes.

Table 6.10
Publicity

Modality	Strategies	Outcome
Print Media	Time-Dependent Articles	Focuses on specific incident, immediacy and danger
Radio and Television	Featured Articles	In-depth description of problem and solution
	Public Service Announcements	Brief, concise verbal message reaching a large audience
	Interviews	Personal opinion and informational exchange
	Talk Shows	Communication of personal experience with an analysis of the problem and solution
Press Conferences	Media Kits	Organized promotion and communication to a variety of media
	Press Releases	Coinciding with the press conference, a brief written description announcing an important event

Table 6.11
Outreach

Activity	Effect
Pamphlets and Brochures	Well-organized, brief written communication with wide distribution to a variety of audiences
Fact Sheets	A general one page description of the problem and solution to be used for multiple purpose
Posters	Visual display and exposure to various target groups
Newsletters	Maintaining regular communication linkages between groups
Speaker's Bureau	Direct education and promotion of program by experienced speakers
Hot Lines	Immediate help and information for those in need of services
Videos	Audio and visual delivery of messages with wide appeal
Web Page	Access to a wide range of information with no time or location

Table 6.10 illustrates how each of these three communication modalities is used to deliver specific types of messages to the public.

The value of a community outreach effort is that very specific populations can be targeted for communication. For example, there may be certain groups of youth living in particular areas of the community whose psychological and social profiles suggest that they are at-risk for becoming involved in firesetting. Or, there may be specific sections of a community where there is a higher incidence of child-set fires, and therefore a greater need to target services. A focused community outreach effort in these areas is likely to be highly effective in reducing the rate of juvenile firesetting. There are many types of outreach activities, from developing and distributing pamphlets and posters to operating a telephone hotline, that can help educate at-risk youth and their families. **Table 6.11** describes many of these activities.

Publicity and community outreach activities can be limited or broad depending on the number of communication modalities employed by program management. At the very minimum, it is recommended that a juvenile firesetter program develop a simple brochure educating parents about the problem of juvenile firesetting and encouraging them to seek help. This brochure can deliver fire prevention and safety information and be distributed to a number of different agencies in the community including schools, hospitals, mental health programs, social services, and the juvenile justice system. The United States Fire Administration (USFA) has developed and is distributing a brochure on juvenile firesetting entitled, *Children and Fire....A Growing Concern*. It is available free of charge by request. Orders for up to 200 copies at a time can be made to the USFA at www.usfa.fema.gov. Educating the community through publicity and outreach is one of the best ways to prevent child-set fires.

Step F. - Continuing Education

To provide state-of-the-art services, juvenile firesetter program management and staff must understand how new information is advancing in the field of juvenile firesetting and arson. They must be offered opportunities to educate themselves with respect to new ideas and emergency trends. There are many ways to provide continuing education experiences. The National Fire Academy has excellent educational programs in juvenile firesetting and arson. Local colleges, some of which offer fire science, can be an important source of information. Many professional organizations, such as the International Association of Fire Chiefs, the International Association of Arson Investigators, and the National Fire Protection Association, all have annual meetings in which juvenile firesetting is often a workshop topic. Several state organizations, such as the Massachusetts Coalition, offer annual educational seminars. Finally, the Internet is rapidly becoming an excellent source of up-to-the-minute information on juvenile firesetting and arson. Relevant Internet addresses are listed in **Appendix 6.4**. The life of a juvenile firesetter

program depends upon its ability to offer the community the highest standard of services based on the most current knowledge and information offered by science and technology.

Summary Points

- **The first task in building a juvenile firesetter program is to determine whether the community has a problem and therefore a need for a program.**

- **If there is a community need, program planning can proceed. This task should result in selecting the site for the juvenile firesetter program, naming the program leader, and specifying the geographic area serviced by the program.**

- **The third task, program development, builds and sets into place all the necessary program operations, including the management team, advisory council, service delivery system, budget, funding, organizational chart, and inter-agency linkages.**

- **Before the doors are open for business, the program must be staffed and the staff trained, documentation procedures must be established, and issues related to confidentiality and liability must be clarified. This marks the fourth task - program implementation.**

- **To ensure the longevity and success of the juvenile firesetter program, short and long range planning activities must be considered such as developing an operations handbook and resource directory, building a monitoring and evaluation system, mounting a public awareness campaign, and providing continuing education opportunities.**

Suggested Reading

Topic	Resource
This is a public awareness program and video on the problem of juvenile firesetting and arson and what steps communities can take to control and prevent it.	Maryland Fire and Rescue Institute (1998). <u>The Faces of Juvenile Fire Setting. A State of Maryland Public Awareness Program.</u> College Park, MD.

Appendix 2.1

Juvenile Miranda Rights Statements

1.   Los Angeles, California

2.   Parker, Colorado

Juvenile Miranda Rights

The following is a set of juvenile Miranda Rights developed by the Los Angeles Grand Jury and recommended for use in the State of California. They are presented here as guidelines. Each fire department must check with its own State to determine the appropriate application of the Miranda Rule.

1. You don't have to talk with us or answer our questions if you don't want to.
2. If you decide to talk with us you have to understand that anything you say can be used against you. We can tell the Probation Officer and the Judge what you tell us.
3. You can talk to a lawyer now if you want to and you can have him with you when we ask our questions.
4. If you want to have a lawyer but you don't have enough money to hire your own, then we will get the judge to get one for you and it won't cost you anything.

Waiver Questions

1. Do you understand what I have said?
2. Do you want to ask me anything?
3. Do you want to talk with me now?
4. Do you want to have a lawyer, or not?

Source:
Los Angeles Grand Jury

PARKER FIRE PROTECTION DISTRICT
ADVISEMENT OF RIGHTS

Name of Person to be Advised_____   Date_____

Place of Advisement_____   Time_____

Agency Case Number_____

I am a Fire Investigator. Before you are asked any questions by a Fire Investigator about crimes involving you, you must understand your rights.

1.   You have the right to remain silent.

2.   Anything you say can be used against you in court.

3.   You have the right to talk with a lawyer before you are questioned, and to have him/her present with you during any questioning.

4.   If you want a lawyer, but cannot afford to hire a lawyer, a lawyer will be appointed by a court to represent you before you are questioned, and be with you during any questioning at no charge to you.

5.   If you decide to start answering questions, you will still have the right to stop answering questions, and also the right to talk to a lawyer at any time.

I have read this statement of my rights. I understand what my rights are. _____

Signature of Person Advised

I read this Advisement of Rights to the person who signed his name above, and I witnessed the making of the above signature.

_____

Signature of Advising Investigator

WAIVER OF RIGHT

I understand my above rights, and I know what I am doing. I agree to answer questions. I do not want a lawyer at this time.

_____

Signature of Person Waiving Rights

Waiver Signature Witnessed by:

_____

Appendix 2.2

Initial Contact Forms

1. **Fire Stoppers**
    King County, Washington

2. **Juvenile Firesetter Prevention Program**
    State of Colorado

3. **Portland Fire and Rescue**
    Portland, Oregon

Fire Stoppers
Incident Referral Form

Incident Number_____   Incident Date_____

Referring Officer: name_____/ employee number_____ ___ ___

Incident Address: _____
                          Street   City   State   Zip

Fire Investigator: _____   Investigator's Incident #_____

Youth Information

Name: _____   Sex M ( ) F ( ) DOB_____

Address: _____
              Street   City   State   Zip

School currently attending: _____   Grade _____

Mother/Guardian: _____

Wk phone (__ __ __)__ __ __-__ __ __ __ home phone: (__ __ __)__ __ __-__ __ __ __

Father/Guardian: _____

Wk phone (__ __ __)__ __ __-__ __ __ __ home phone: (__ __ __) __ __ __-__ __ __ __

Where did the incident/fire occur? _____

Items ignited: _____

Source of ignition: matches ( )   lighter ( )   other ( )

Others involved in incident   yes ( ) *list names on reverse side of this form*   no ( )

When applicable

Were smoke detectors present?

Did they activate? Yes ( )   No ( ) (if no why) _____

(When appropriate, test all smoke detectors and provide a new detector/battery)

If matches and lighters are accessable to children please ask parent/caregivers to remove them immediately. You will want to explain some about our program and that the parent/guardian can expect a call from the Prevention Division to extend these services and explain the intervention program in greater detail.

Comments

COLORADO JUVENILE FIRESETTER PREVENTION PROGRAM CONTACT FORM

_____ **DEPT. NAME** _____ **Inc. Census Tract** _____ **County**

INCIDENT DATE: _____ NO. _____ TIME _____ CR NO. _____
INCIDENT ADDRESS: _____ Street _____ City _____ Zip

Multiple Juveniles ☐ Y ☐ N # _____ Ignition Source: ☐ Match ☐ Lighter ☐ Other
☐ Flammable Liquid/Accelerant Used

Loss: $ _____ Intentional: ☐ Y ☐ N Injuries: ☐ Y ☐ N # _____ Deaths: ☐ Y ☐ N # _____
Hospitalizations: ☐ Y ☐ N # _____ Describe Injuries/Deaths _____

Location of Fire: Outside-Location of Origin _____ ☐ Inside / ☐ Inside Occupied Room of Origin _____

Referral Source Name: _____ Agency/Address: _____ Phone: _____
☐ Care giver ☐ School ☐ Law Enforcement ☐ Mental Health ☐ Fire Service ☐ Juvenile Justice
☐ Parent ☐ Other/Describe _____

Care giver/Parent Smokes ☐ Y ☐ N Did the home meet community standards for health/welfare of the
child? ☐ Y ☐ N
Was the child supervised by a person 12 years of age or older at the time of the incident? ☐ Y ☐ N

Description of the Incident and Pertinent Information:
_____

Report by: _____ _____
Printed Name Signature

Juvenile Information

Last Name: _____ First Name: _____ M.I. _____ DOB ___/___/___
Sex ☐ M ☐ F Race: ☐ White ☐ Asian ☐ African Am. ☐ Native Am. ☐ Hispanic ☐ Other
Age: _____ Grade in School _____ School Currently Attending_____
Soc. Sec. #: _____-_____-_____

Home Address: _____ Phone: _____

Adult No. 1 Residing With The Child

Name:_____
Address: _____
Phone: H _____ W _____
Employed: ☐ Y ☐ N
Marital Status: ☐ Married ☐ Separated
☐ Divorced ☐ Remarried ☐ Widowed
Relation to Juvenile: ☐ Natural ☐ Step
☐ Adoptive ☐ Foster ☐ Grandparent
☐ Other _____

Adult No. 2 Residing With The Child

Name: _____
Address: _____
Phone: H _____ W _____
Employed: ☐ Y ☐ N
Marital Status: ☐ Married ☐ Separated
☐ Divorced ☐ Remarried ☐ Widowed
Relation to Juvenile: ☐ Natural ☐ Step
☐ Adoptive ☐ Foster ☐ Grandparent
☐ Other _____

Others Residing With The Child

Name: _____   Relationship: _____
Name: _____   Relationship: _____
Name: _____   Relationship: _____
Name: _____   Relationship: _____

JFS CONTACT RECORD

Date _____   Time _____

TYPE OF CONTACT: _____
                                      (contact name)

Telephone _____   In Person _____
    No contact _____   appt. scheduled _____
    message: call us _____   education/assessment _____
    message: we'll call _____   education only _____
    phone disconnected _____   referral _____
    appt. scheduled _____   other _____
    referral by phone _____
    other _____

CONTACT INITIATED BY: PFB _____   CLIENT _____   OTHER _____

COMMENTS (SEE ATTACHED NARRATIVE):

_____
_____
_____
_____

****

JFS CONTACT RECORD

Date _____   Time _____

TYPE OF CONTACT: _____
                                      (contact name)

Telephone _____   In Person _____
    No contact _____   appt. scheduled _____
    message: call us _____   education/assessment _____
    message: we'll call _____   education only _____
    phone disconnected _____   referral _____
    appt. scheduled _____   other _____
    referral by phone _____
    other _____

CONTACT INITIATED BY: PFB _____   CLIENT _____   OTHER _____

COMMENTS (SEE ATTACHED NARRATIVE):

_____
_____
_____
_____

Portland's Juvenile Firesetter Contact Record

JFS CONTACT RECORD

Date _____   Time _____

TYPE OF CONTACT: _____
                                                          (contact name)

Telephone _____   In Person _____
    No contact _____   appt. scheduled _____
      message: call us _____   education/assessment _____
   message: we'll call _____   education only _____
   phone disconnected _____   referral _____
    appt. scheduled _____   other _____
   referral by phone _____
      other _____

CONTACT INITIATED BY: PFB _____   CLIENT _____   OTHER _____

COMMENTS (SEE ATTACHED NARRATIVE):

_____
_____
_____
_____

****

NARRATIVE:

_____

_____

_____

_____

(CONTINUE ON BACK IF NECESSARY)

Portland's Juvenile Firesetter Contact Record

Appendix 3.1

Juvenile Firesetter Child and Family Risk Surveys

Juvenile Firesetter Child and Family Risk Surveys
Description and Instructions
Colorado Juvenile Firesetter Prevention Program

Survey Development

In September 1995, the Colorado Department of Public Safety/Division of Fire Safety was awarded a grant to design and test the applicability and effectiveness of the Juvenile Firesetter/Arson Control and Prevention Program model for statewide dissemination. Funding for this program was provided by the Federal Emergency Management Agency, U.S. Fire Administration (EMW-95-S-4780), under P.L. 103-254, the Federal Arson Prevention Act of 1994. Also, the Adam and Dorothy Miller Lifesafety Center, Inc. (dba Miller Safety Center) was awarded a grant in 1991 to develop a pilot program based upon the model produced by the Institute for Social Analysis for the Bureau of Justice, Office of Juvenile Justice and Delinquency Prevention and the U.S. Fire Administration under Cooperative Agreement #JN-CX-K002, "The National Juvenile Justice Firesetter/Arson Control and Prevention Program."

The Miller Safety Center determined that the fire service needed a risk assessment tool that was accurate for predicting future risk of firesetting in juveniles, yet offered a reduction in the length of time needed to conduct the evaluation. The Colorado Project's primary objective was to develop a juvenile fire risk survey for the fire service. Kenneth Fineman, Ph.D., the primary author of the U.S. Fire Administration's juvenile firesetter evaluation which was first published in the 1970's and updated throughout the 1980's, offered his most current, unpublished version of this instrument as the basis for the Colorado Project. In the fall of 1995, Fineman and members of the Colorado Project (Marion Doctor, LCSW; Joe B. Day; Larry Marshburn; Kenneth Rester, Jr.; Cheryl Poage; Paul Cooke; Carmen Velasquez; Michael Moynihan, Ph.D., and Elise Flesher, Ph.D. candidate), met to revise the juvenile firesetter evaluation so that it could be used for research purposes. The result was the Comprehensive FireRisk Assessment, published in the Colorado Juvenile Firesetter Prevention Program. Training Seminar. Volume 1.

In 1998, using the Comprehensive FireRisk Assessment, Moynihan and Flesher conducted a study to develop the Juvenile Firesetter Child and Family Risk Surveys. The method and results of this study are reported in detail in their research paper (1998) cited in the reference list. From the Comprehensive FireRisk Assessment, Moynihan and Flesher identified a subset of statistically valid questions to comprise the Risk Surveys. Hence, the questions on the Risk Surveys are derived directly from the questions on the Comprehensive FireRisk Assessment. The Risk Surveys represent a shortened version of the Comprehensive FireRisk Assessment.

Survey Use

The Juvenile Firesetter Child and Family Risk Surveys offer an accurate means to assess the risk of future firesetting in juveniles. They are comprised of two sections, the Child Risk Survey (for the juvenile) and the Family Risk Survey (for the parent). The Risk Surveys take about thirty minutes to administer. It is recommended that the Risk Surveys be conducted in an interview format with the juvenile and at least one parent. The Risk Surveys do not release the fire service from the need to properly conduct cause and origin investigations, case documentation, obtain proper parental releases to interview a child, network community referral resources, and provide intervention education when appropriate.

When using the Risk Surveys, the following procedures are recommended.
- **Develop rapport with the family.**
- **Explain to the juvenile and parents the purpose of the interview.**
- **Obtain written permission from the parent or legal guardian to conduct the Child Survey.**
- **Complete all the demographic information.**
- **First conduct the Family Survey without the child present.**

- If possible, conduct the Child Survey without the parents present in the same room.
- Begin the Child Survey with the Development of Rapport section.
- Ask all the questions exactly as they are written, to conform to the validated protocol.

It is also recommended that both the Family and Child Surveys be conducted. The highest degree of accuracy will be achieved if both surveys are used. The Family Survey can be conducted over the phone with the child's parent; however, the Child Survey must be conducted in person and only after the proper parental release has been signed. It is also recommended that a fire or police incident report be placed in the file whenever possible.

While the questions on the Child and Family Surveys must be asked as they are written, there may be circumstances in individual cases where additional information is obtained. Please be sure to write notes in the case file regarding any information that is offered during the interview, even if the information is not scored.

Survey Scoring

Total the numerical weights assigned to the answers received during the interview. The following table shows how the total scores on the Child and Family Surveys correspond to the levels of firesetting risk and related methods of intervention.

Risk Level	Source	Score	Intervention
Little	Family Survey	<429	Education
Little	Child Survey	<511	Education
Definite	Family Survey	429<457	Referral and Education
Definite	Child Survey	511<540	Referral and Education

If the Child Risk Score is equal to or greater than 511, but less than 540, and/or the Family Risk Score is equal to or greater than 429, but less than 457 consider conducting the Comprehensive FireRisk Evaluation both the child and the parents or refer to a mental health professional.

Extreme	Family Survey	>457	Referral
Extreme	Child Survey	>540	Referral

There are discretionary areas where it may be advisable to conduct the Comprehensive FireRisk Evaluation initially. The comprehensive FireRisk Evaluation is recommended for cases which may involve the following factors.

- When the family is referred by social services, mental health, probation, or in some cases, juvenile diversion.
- When a resistant or uncooperative child or parent has been encountered.

References

Moynihan, M. and Flesher, E. Locating a Risk Cut-Off Level Based on Key Variables in the Regression Equation. Child Interview. Parent Interview. Boulder, CO: Department of Psychology, University of Colorado, 1998.

Poage, C., Doctor, M., Day, J.B., Rester, K., Velasquez, C., Moynihan, M., Flesher, E., Cooke, P., Marshburn, L. (1997). Colorado Juvenile Firesetter Prevention Program: Training Seminar. Vol. 1. Denver, CO: Colorado Division of Firesafety.

PARTICIPATION RELEASE

The _____ utilizes the juvenile firesetter screening program developed by the Federal Emergency Management Agency and the United States Fire Administration to evaluate the child that has been involved in a fire incident or has been referred to the City by a parent or another entity or agency.

The evaluation tries to assess the risk of involvement in future firesetting behavior. To do this, six areas describing individual characteristics are evaluated (demographic, physical, cognitive, emotional, motivation, and psychiatric).

Based on the results of the evaluation, your child's tendencies will place him/her in one of the following areas of concern:

Little Risk   -   needs educational intervention

Definite Risk   -   needs referral for evaluation to a mental health agency or to
                                      a licensed psychologist or psychiatrist and educational intervention

Extreme Risk   -   needs immediate referral for evaluation
                                      by a licensed psychologist or psychiatrist

If educational intervention is indicated, the _____ program will offer further educational activity for your child.

Depending on the circumstances regarding an individual case, other agencies such as the school your child attends, local law enforcement, social services departments, etc. may become involved.

The questions asked in this evaluation may be viewed prior to signing this release upon request.

I, _____, have read the previous statement and do hereby grant permission for my child, _____, to participate in the _____ Intervention Program and hereby authorize to release information regarding my child to such other governmental entities and agencies as it may deem appropriate.

_____   _____
            Parent/Guardian   Date/Time

_____   _____
                  Juvenile   Witness

COMPREHENSIVE FAMILY FIRERISK INTERVIEW FORM
(Questions to be asked of Parents of Children 3 to 18 Years of Age)

JUVENILE FIRESETTER CONTACT FORM _____DEPT. NAME _____Inc. Census Tract____County

INCIDENT-DATE_____ NO. _____ TIME_____ CR. NO_____
INCIDENT ADDRESS:_____Street _____City _____ZIP

Multiple Juveniles ☐ Y ☐ N #_____ Ignition Source: ☐ Match ☐ Lighter ☐ Other ☐ Flammable Liquid/Accelerant Used

Est. Loss: $_____ Intentional: ☐ Y ☐ N Injuries: ☐ Y ☐ N #____ Death: ☐ Y ☐ N #_____
Hospitalizations: ☐Y ☐N #_____ Describe Injuries/Deaths _____

Location of Fire: Outside-Location of Origin _____ ☐ Inside/☐ Inside-Occupied Room of Origin_____

Referral Source Name:_____ Agency/Address: _____ Phone: _____
☐ Caregiver ☐ School ☐ Law Enforcement ☐ Mental Health ☐ Fire Service ☐ Juvenile Justice
☐ Parent ☐ Other/Describe _____

Caregiver/Parent Smokes ☐ Y ☐ N Did the home meet community standards for health/welfare of the child? ☐ Y ☐ N

Was the child supervised by a person 12 years of age or older at the time of the incident? ☐ Y ☐ N

Description of Incident and Pertinent Information:
_____
_____

Report by: _____ _____
Printed Name Signature

Juvenile Information

Last Name: _____ First Name: _____ M.I. _____ DOB _____/____/____
Sex ☐ M ☐ F Race: ☐ White ☐ Asian ☐ African Am. ☐ Native Am. ☐ Hispanic ☐ Other
Age: _____ Grade in School_____ School Currently Attending _____
Soc. Sec. #: _____-____-____

Home Address: _____ Phone: _____

Adult No. 1 Residing With The Child
Name:_____

Address: _____

Phone: H _____ W _____
Employed: ☐ Y ☐ N
Marital Status: ☐ Married ☐ Separated ☐ Divorced ☐ Remarried ☐ Widowed

Relation to Juvenile: ☐ Natural ☐ Step

Adult No. 2 Residing With The Child
Name:_____

Address: _____

Phone: H _____ W _____
Employed: ☐ Y ☐ N
Marital Status: ☐ Married ☐ Separated ☐ Divorced ☐ Remarried ☐ Widowed

Relation to Juvenile: ☐ Natural ☐ Step

Others Residing With The Child
Name: _____ Relationship: _____
Name: _____ Relationship: _____
Name: _____ Relationship: _____
Name: _____ Relationship: _____

Moynihan, Flesher, and Colorado Juvenile Firesetter Prevention Program Staff   06/29/98   Family Risk Survey

JUVENILE FIRESETTER FAMILY RISK SURVEY   Date Survey Conducted: _____

This Family Risk Survey is designed to be given to parents who have concerns about their child's fire play or firesetting behavior or whose child has set a fire which has come to the attention of a fire department, police agency or other community agencies. The Family Risk Survey is intended for use only as a preliminary screening tool and should be used with the Child Risk Survey to assess the child's suitability for fire intervention education or mental health referral.

The Family Risk Survey may be administered to parents over the phone or in person. The Child Risk Survey should be administered to the child in person and separate from their parents only after the parents or guardians have provided written informed consent for the child's participation in the survey.

Prior to administering the Family Risk Survey, please provide the following incident and demographic information.

I. Incident #: _____ Incident Date: ____/ ____/ ____ Incident Location: _____ CR #: _____

Incident Description: _____

_____

_____

_____

II. Child's Last Name:_____ First Name: _____ M.I. _____ D.O.B. ___/___/___

Child's Address: _____ Home Phone: _____

School Child Attends: _____ Grade: _____

III. Name of Parent/Guardian providing information: _____

Address if different from Child's: _____ Work Phone: _____

IV. Referral Source if **not** a fire call (Name/Agency): _____

Agency's Address: _____ Phone: _____

V. Interviewer's Name: _____ Phone: _____

Interviewer's Affiliation: _____

Interviewer's notes and/or comments: _____

_____

_____

_____

_____

Moynihan, Flesher, and Colorado Juvenile Firesetter Prevention Program Staff   06/29/98   Family Risk Survey
*Original questions appear in Fineman, (1996), Comprehensive Fire Risk Assessment, Published in the Colorado Juvenile Firesetter Prevention Program: Training Seminar Vol. I, (1997).

JUVENILE FIRESETTER FAMILY RISK SURVEY   Date Survey Conducted: _____

To administer: Ask the question as written, check the response, place the appropriate constant weight in the score column, and add the scores to determine the Total Family Risk Score. Please substitute the child's name in questions 1-5.

Questions*	Constant	Score

1. If you had to describe (child's name) curiosity about fire, would you say it was absent, mild, moderate, or extreme?

		Constant	Score
absent	_____	0	_____
mild	_____	99	_____
moderate	_____	198	_____
extreme	_____	297	_____

2. Has (child's name) been diagnosed with any impulse control conditions, such as Attention Deficit Disorder (ADD) or Attention Deficit Disorder with Hyperactivity (ADHD)?

yes	_____	_____(Diagnosis)	28	_____
no	_____		0	_____

3. Has (child's name) been in trouble outside of school for non-fire related behavior?

yes	_____	_____(What?)	90	_____
no	_____		0	_____

4. Has (child's name) ever stolen or shoplifted?

yes	_____	14	_____
no	_____	0	_____
dk/na	_____	0	_____

5. Has (child's name) ever beat up or hurt others?

yes	_____	14	_____
no	_____	0	_____
dk/na	_____	0	_____

6. Besides this fireplay or firesetting incident, how many other times has your child played with fire, including matches or lighters, or set something on fire?

1 (current)	_____	84	_____
2 (current + 1)	_____	168	_____
4 (current + 2-4)	_____	336	_____
6 (current + 5)	_____	504	_____

7. Is there an impulsive (sudden urge) quality to your child's firesetting or fire play?

yes	_____	71	_____
no	_____	0	_____
dk/na	_____	0	_____

TOTAL FAMILY RISK SCORE _____

Question (8) is for informational purposes and does not score.

8. *Is there a history of emotional, physical, or sexual abuse in the family?   Yes _____ No _____*
   Who _____   Relationship _____   Currently in the home _____
   If there are indications of abuse or neglect, consult with social services or law enforcement immediately

A. **The Cut Off Score For Mental Health Referral For the Family Risk Survey Is 457 or Above.** If either the Family Risk Survey is equal to or greater than 457 and/or the Child Risk Survey is equal to or greater than 540, the child should be referred to a mental health professional.

B. If either the Family Risk Score is equal to or greater than, 429, but less than 457 and/or the Child Risk Score is equal to or greater 511, but less than 540 consider conducting the comprehensive firesetter risk assessments for both the child and the parents or refer to a mental health professional.

C. **AN INTERVENTION EDUCATION PROGRAM** is appropriate if the Family Risk Score is less than 429 and/or the Child Risk Score is less than 511.

Moynihan, Flesher, and Colorado Juvenile Firesetter Prevention Program Staff   06/29/98   Family Risk Survey
*Original questions appear in Fineman, *(1996), Comprehensive Fire Risk Assessment*, Published in the Colorado Juvenile Firesetter Prevention Program: Training Seminar Vol. I, (1997).

JUVENILE FIRESETTER CHILD RISK SURVEY   Date Survey Conducted: _____

This Child Risk Survey is designed to be given to children (with their parent's written informed consent) who have played with fire or who have set a fire which has come to the attention of a fire department, police agency or other community agencies. The Child Risk Survey is intended for use only as a preliminary screening tool and should be used with the Family Risk Survey to assess the child's suitability for fire intervention education or mental health referral.

The Family Risk Survey may be administered to parents over the phone or in person. The Child Risk Survey should be administered to the child, in person, and separate from their parents only after the parents or guardians have provided written informed consent for the child's participation in the survey.

Prior to administering the Child Risk Survey, please provide the following incident and demographic information if it has not already been provided in the Family Risk Survey section.

I. Incident #: _____   Incident Date: ____/____/____   Incident Location: _____   CR #: _____

Incident Description: _____

_____

_____

_____

II. Child's Last Name: _____   First Name: _____   M.I. _____   D.O.B. __/__/__

Child's Address: _____   Home Phone: _____

School Child Attends: _____   Grade: _____

III. Name of Parent/Guardian providing information: _____

Address if different from Child's: _____   Work Phone: _____

IV. Referral Source if **not** a fire call (Name/Agency): _____

Agency's Address: _____   Phone: _____

V. Surveyor's Name: _____   Phone: _____

Surveyor's Affiliation: _____

Surveyor's notes and/or comments: _____

_____

_____

_____

JUVENILE FIRESETTER CHILD RISK SURVEY   Date Survey Conducted: _____

INFORMATIONAL ACTIVITY FOR THE CHILD

Have the child draw a picture of the fire or fireplay incident and/or write a paragraph describing why they are in your office today while you are conducting the Family Survey with the parents.

DEVELOPMENT OF RAPPORT

The purpose of this section is to make the child comfortable with you. The more at ease you can make him, the greater the likelihood that he will answer all of your questions. If the following questions aren't enough, add your own. Questions or language can be modified in the Development of Rapport section only, **all other questions should be asked as written.** This section was developed by Kenneth R. Fineman Ph.D., and is reprinted from Comprehensive Fire Risk Assessment as published in the Colorado Juvenile Firesetter Prevention Program: Training Seminar Vol. I.

1. [Introduce yourself] I'm _____   What's your name?_____

2. How old are you? _____

3. What school do you go to? _____   What grade are you in? _____

    Do you like your school?_____   Are there nice/okay teachers at your school? _____

4. What classes/subjects do you like/not like? _____

5. What do you do for fun? Do you have hobbies? _____

6. Who's your best friend? _____

7. What do you like to play/do with your friend? _____

8. What do you watch on TV and/or what videos do you watch? _____

9. What is your favorite person/show on TV? _____

10. What is your favorite video/computer game? _____

11. What do you like about that game? [Is there extreme interest in violence or fire?] _____

[When rapport is established, determine level of understanding if the child is under 7 or appears to have problems communicating.]

COMPARISON OF THE ORIGINAL AND REVERSE ORDER VERSIONS OF THE INCIDENT

For children age nine and older, consider asking the following prior to proceeding:

Have the child describe their involvement in the incident from some point in time prior to some point in time after the incident. At the end of the interview ask the child to repeat this description in reverse order.

The average child whom is at least nine years old should be able to relate incident details in reverse order if the original version of his/her account of the incident was truthful. If the order of events is significantly different in the order.

<u>JUVENILE FIRESETTER CHILD RISK SURVEY</u>   Date Survey Conducted:_____

DETERMINE LEVEL OF UNDERSTANDING (Under 7)

<u>This section was developed by Kenneth R. Fineman, Ph.D., and is reprinted from the
Comprehensive Fire Risk Assessment as published in the Colorado Juvenile Firesetter
Prevention Program: Training Seminar Vol. I.</u>

It is often difficult to determine if a young child really understands you. (This section may be
skipped if you are interviewing an older child). There may be an age barrier, a language barrier,
a learning problem, or sub-normal intelligence. It is fruitless to go through an entire interview
unless you are first assured that the child has enough understanding to complete the interview.
There are several ways to gauge whether you are on the same "wave length" as the child. The
following are suggested ways to do so:

a. Obtain information from rapport section above:
     By paying close attention to the manner in which a young child responds to the 11
     questions above, you can estimate whether he can understand and respond to the other
     questions in this instrument.

b. Using crayons/paper as a tool:
     You can ask the child to draw pictures of common objects, his favorite toys, houses,
     trees, and people. Then ask him to describe what he has drawn. Clear explanations of his
     drawings and the action taking place in some of those drawings will tell you something
     about the child's vocabulary and his ability to understand.

c. Using toys and games:
     Have toys of the appropriate developmental level of the child available. Engage the child
     in a game with the toys or allow the child free play with the toys. After a while ask the
     child about the toys and the game he is playing. Inquire about the rules, the purpose, etc.
     Estimate the child's vocabulary in terms of his ability to complete the interview.

d. Using puppets:
     Have hand puppets available. Allow the child to set the interaction, with the child
     playing all parts or with you playing some of the parts. Quiet children can become quite
     verbal with this approach. Focus on the child's ability to understand your questions
     during the puppet play and determine if this level of communication is sufficient for
     continued interviewing.

If you are satisfied that the child has adequate understanding, proceed with the interview.

JUVENILE FIRESETTER CHILD RISK SURVEY   Date Survey Conducted _____

To administer: Ask the question as written, check the response, place the appropriate constant weight in the score column, and add the scores to determine the Total Child Risk Score.

Questions*		Constant	Score
1. Do you have any brothers or sisters?			
yes _____		0	_____
no _____ **(If no, skip to Q. 3)**		0	_____
2. How well do you get along with them?			
always get along _____		28	_____
Score only one usually get along _____		56	_____
response, using sometimes get along _____		84	_____
the one with the don't get along very often _____		112	_____
highest risk value. never get along _____		140	_____
3. How well do you get along with your mother?			
always get along _____		10.5	_____
usually get along _____		21	_____
sometimes get along _____		31.5	_____
don't get along very often _____		42	_____
never get along _____		52.5	_____
4. Do you fight or argue with your mother?			
never _____		10.5	_____
rarely _____		21	_____
sometimes _____		31.5	_____
usually _____		42	_____
always _____		52.5	_____
5. Do you see your father as much as you'd like?			
yes _____		0	_____
no _____		60	_____
too much _____		60	_____
6. When you are asked to do something, do you usually do it?			
yes _____		0	_____
no _____		17.5	_____
7. Do you lie a lot?			
yes _____		17.5	_____
no _____		0	_____

8. What happens at home when you get in trouble?

grounded _____	physical punishment _____	0.0	_____
talked/lectured _____	sought outside help _____	0.0	_____
abused** _____	other/nothing _____	0.0	_____
	yelled at _____	32	_____

9. Has there been an ongoing (chronic) crisis or problem in your life or in your family?

yes _____ _____ (What?)	62	_____
no _____	0	_____

Moynihan, Flesher, and Colorado Juvenile Firesetter Prevention Program Staff   06/29/98   Child Risk Survey
*Original questions appear in Fineman, *(1996), Comprehensive Fire Risk Assessment*, Published in the Colorado Juvenile Firesetter Prevention Program: Training Seminar Vol. 1, (1997).

JUVENILE FIRESETTER CHILD RISK SURVEY   Date Survey Conducted _____

10. Besides this fireplay or firesetting incident, how many other times have you played with fire, including matches or lighters, or set something on fire?

1 (current)	_____	32	_____
2 (current +1)	_____	64	_____
4 (current +2-4)	_____	128	_____
6 (current +5)	_____	192	_____

11. What did you do after the fire started?

put it out	_____	called for help	_____	0,0	_____
ran away	_____	didn't try to run	_____	0,0	_____
panicked	_____	tried to extinguish	_____	0,0	_____
other	_____	didn't try to extinguish	_____	0,0	_____
		stayed and watched	_____	40	_____

12. Did you intend to play with fire or set the fire, that is, did you play with or set the fire on purpose?

yes	_____	187	_____
no	_____	0	_____

If the surveyor has evidence of intent, the surveyor may override the youth's denial

13. Where did you set the fire?

_____

(If any type of structure was involved as a target or a location, score:)   47 _____   _____
0 _____

other   _____

14. Do you like to look at fire for long periods of time?

yes	_____	250	_____
no	_____	0	_____

TOTAL CHILD RISK SCORE _____

Question (15) is for informational purposes and does not score.

15. *How did you get the ignition source (match/light/other) used in the fire/fireplay?*

_____

** If there are indications of abuse or neglect consult with social services or law enforcement immediately.

If the child is at least nine years old, ask the child to repeat, in reverse order, the description of the incident.
How does this compare to the original description?

A. **The Cut Off Score For Mental Health Referral For The Child Risk Survey Is 540 or Above.** If either the Child Risk Survey is equal to or greater than 540 and/or the Family Risk Survey is equal to or greater than 457, the child should be referred to a mental health professional.

B. If the Child Risk Score is equal to or greater than 511, but less than 540, and/or the Family Risk Score is equal to or greater 429, but less than 457 consider conducting the comprehensive firesetter risk assessments for both the child and the parents or refer to a mental health professional.

C. **AN INTERVENTION EDUCATION PROGRAM** is appropriate if the Child Risk Score is less than 511 and/or the Family Risk Score is less than 429.

Moynihan, Flesher, and Colorado Juvenile Firesetter Prevention Program Staff   06/29/98   Child Risk Survey
*Original questions appear in Fineman, *(1996), Comprehensive Fire Risk Assessment*, Published in the Colorado Juvenile Firesetter Prevention Program: Training Seminar Vol. 1, (1997).

RELEASE OF LIABILITY

I do hereby release, indemnify, and hold harmless the _____
Juvenile Firesetter Intervention Program, all its employees and volunteers against all claims, suits, or actions of any kind and nature whatsoever which are brought or which may be brought against the _____ Juvenile Firesetter Intervention Program for, or as a result of any injuries from, participation in this program.

_____   _____
Parent/Guardian   Date/Time

_____   _____
Juvenile   Witness

RELEASE OF CONFIDENTIAL INFORMATION

Juvenile's Name _____   D.O.B. _____

Release to/Exchange with:

    Name _____

    Address _____

              _____

    Phone _____

Information Requested _____

_____

I consent to a release of information to and/or an exchange of information with the _____ Juvenile Firesetter Intervention Program. I understand that this consent may include disclosure of material that is protected by state law and/or federal regulations applicable to either mental health or drug/alcohol abuse or both.

This form does not authorize re-disclosure of medical information beyond the limits of this consent. Where information has been disclosed from records protected by Federal Law for drug/alcohol abuse records or by State Law for mental health records, federal requirements prohibit further disclosure without the specific written consent of the patient. A general authorization for release of medical or other information is not sufficient for these purposes. Civil and/or criminal penalties may attach for unauthorized disclosure of drug/alcohol abuse or mental health information.

A copy of this Release shall be as valid as the original.

_____   _____
          Parent/Guardian   Date/Time

_____   _____
             Juvenile   Witness

RISK ADVISEMENT

I have been informed that the FEMA/USFA Juvenile Firesetter Evaluation indicates that my child, _____ has a serious risk of continued involvement with fire setting activity.

I have also been informed by the _____ Juvenile Firesetter Intervention Program of the serious risk of injury and property damage that may continue to exist until the problem is resolved.

I have been advised to seek an evaluation by a licensed psychotherapist or psychiatrist.

_____   _____
Parent/Guardian   Date/Time

_____
Witness

Appendix 3.2

Comprehensive FireRisk Evaluation

1. Instructions
2. Family Evaluation
3. Child Evaluation
4. Parent Questionnaire (English)
5. Parent Questionnaire (Spanish)
6. Parent Questionnaire Visual Scoring Sheet
7. The Structured Category Profile Sheet
8. Participation Release
9. Release of Liability
10. Release of Confidential Information
11. Risk Advisement

INSTRUCTIONS FOR USING THE COMPREHENSIVE FIRERISK FAMILY AND CHILD EVALUATION
Kenneth R. Fineman, Ph.D.

General Instructions

The Comprehensive FireRisk Evaluation was developed to help you acquire the information you need to determine risk; specifically, the determination of little risk, definite risk, or extreme risk, relative to the prediction of future firesetting, and especially dangerous firesetting. To accomplish this you must have a child or family member answer your questions honestly and completely.

The parent questionnaire and the child and family interview forms are constructed so you can score most responses as C-1, C-2, C-3, P-1, P-2, or P-3. A C-2 or -3, or a P-2 or -3 response suggests that the child or parent answered in a way consistent with those who are pathological firesetters or recidivist firesetters. C-2 or -3, or P-2 or -3 responses may also suggest the presence of emotional or behavioral dysfunction. Positioning a C or P response in column 2 of a 3 column matrix indicates definite risk for further and dangerous firesetting. Positioning a C or P in column 3 suggests extreme risk (due either to the child's focus on fire, the likelihood of emotional or behavioral dysfunction, or both).

When a child is given a C-1 or a parent is given a P-1, this indicates that the child or parent is engaging in a behavior that is quite normal or a behavior that is indicative of curiosity firesetting and is not correlated with recidivistic firesetting. It is important that a C-1 or P-1 not be assigned without good reason; since doing so signifies the normalcy of a response. If a response is not normal and it is assigned a C-1 or P-1, the statistics upon which prediction of risk is based becomes distorted.

Some questions are for general information only and are not scored. Some are geared toward setting the groundwork for the questions to follow that are scored. Sometimes there will be many responses that are correct. When this happens mark all that are accurate. However, when it comes time to score the response on the profile sheet, only score (i.e., give credit for) the most severe response. When narrative information is required and you run out of room, use the back of the form.

For some questions you are offered the option of a C-1, C-2, or C-3, and/or a P-1, P-2, or P-3 response. When offered only C responses to choose from, only one C response is required. (In other words, it's either a C-1, a C-2, or a C-3.) When offered only P responses, only one P response is required (P-1, P-2, or P-3). However, when given an option such as C-1 or C-2 or C-3 or P-1 or P-2 or P-3, you are given the opportunity to choose two responses, one from each category. You may also choose only one response, from either the C category or the P category. It is only appropriate to choose two responses, one from each category, if the answer to an item suggests some degree of concern for both the child (C) and the parent (P) or family (P).

Fineman, K. (1996). Comprehensive Fire Risk Assessment Instructions

When Opposite Responses Can Both Get a C-1 or P-1

It is important to think of a C-1 or P-1 response as signifying appropriateness, and C-2 or -3 and P-2 or -3 responses as signifying inappropriateness. By this we mean that the choice of one response over the other must be thought of in terms of the overall context in which the child lives and functions.

As an example, spending what appears to be enough time with a child, while usually being scored a P-1 may actually require a P-2 if the child is being ill treated by the parent. A child staying to watch a fire, or choosing to run away (seemingly opposite responses) can both generate a C-1 if you judge that those behaviors are appropriate responses under the circumstances that you uncover.

Clarifying Your Choices

As an interviewer, you have the option to obtain more information on any question when you feel it is necessary to help you make you C-1/-2/-3 and/or P-1/-2/-3 decision. Within the limits of the time you can allow for an interview, the more information you get the better. Also, when you choose to give a C or P based on a parent of child's "other" response, please elaborate on what "other" means for greater clarification in the future. When you are unsure if a response falls more into a column 1 vs. 2, or a column 2 vs. 3, have the interviewee explain his answer.

If a child is being home schooled, answer only questions 1, 3, and 4 on the child interview and evaluation form, in the school section.

When you answer questions that deal with whether a structure was or was not occupied at the time of the fire, answer the question in terms of what was actually set on fire as opposed to what the juvenile says he intended. As an example, an occupied structure is one that had people in it at the time of the firestart, an unoccupied structure is unoccupied if it had no one in it at the time of the firestart, even if it usually does. A vacant structure is one that not only did not have occupation at the time of the fire, but is generally believed not to, such as a structure in the process of being built.

When answering questions concerning where a child got his firesetting material, consider the most appropriate answer, not the most obvious. Thus, determine the sequence of how the child got his matches before deciding on the response to circle.

Clarifying the Child or Family's Choices

If after you have asked the question exactly as it is written, you feel that the child or parent does not understand a question, either because of the way it is phrased or because they don't understand a word; you have the option to change the way the question is stated to make it clear to the child or parent. You also have the option to substitute a word to be understood.

In order that the questionnaires be applicable to all ages it has been necessary to insert optional language. As an example, you might want to talk to a younger child about his teacher, but to an older child about his classes or subjects. Thus a question may give you a choice of words such as teacher/subject and it is up to you to use the correct word or phrase depending upon the age of the child.

The Format of the Interview Forms and the Parent Questionnaire

Both the original assessment tools in the FEMA manuals as well as the present updated tools are based on the dynamic-behavior theory of firesetting (Fineman, 1980, 1995). The original forms were less structured and less complex. The present forms have greater structure and at the same time provide wider latitude for the fire evaluator to explore the factors that lead to higher risk for future firesetting. The dynamic-behavioral model suggests that past history of dysfunctional behavior coupled with poor supervision and training in fire safety generates an at-risk child. Add to this a traumatic event to lessen the child's inhibitions and increase his impulsiveness, and we are poised for a firestart.

The model further suggests that certain thoughts and feelings that occur before, during, and after the fire should be investigated, as that information will help us understand the motivation for the firesetting and provide very specific information for the referral source who will provide the therapy for those assessed as definite and extreme risk. The present instruments are set up in such a manner as to allow the evaluator to more clearly understand the sequence of thoughts, feelings, and behavior that lead to and maintain firesetting.

You may use the number of column 2 or 3 responses on each of the three instruments, or their additive value as represented on the structure category profile sheet, to understand the sequence as well as to assess risk. Probably the easiest method will be to calculate the percentages on the forms, as discussed below.

On some occasions you may not be able to interview the family, as only the child will be available for the interview. In those situations, use the first sheet of the family interview form with the child in order to get as much information about the family and living arrangements as possible.

The Child Evaluation

This interview form is divided into eight content sections plus demographics. As you interview, circle C or P responses and write in narrative information that you want to remember. When the interview is completed, count up all C-1 responses and enter that number in the appropriate square on the small summary box that is included at the end of each of the eight sections. Repeat this process for C-2 through P-3. When complete, transfer that information to the large summary box at the end of the interview form. Then total each column and record that sum in the appropriate square. Once you have all totals recorded, use the total score for each of the columns to calculate the percentage of risk for child, family, and total risk according to the following formula.

Child Risk $\quad \dfrac{C2+C3}{C1+C2+C3} \quad = \underline{\hspace{2cm}} \%$

Family Risk $\quad \dfrac{P2+P3}{P1+P2+P3} \quad = \underline{\hspace{2cm}} \%$

Total Risk $\quad \dfrac{C2+P2+C3+P3}{C1+P1+C2+P2+C3+P3} \quad = \underline{\hspace{2cm}} \%$

Does a child see fire as having special, miraculous, or spiritual powers? And if he does, how do we know if it's a C-2 or C-3 response? The evaluation that you are conducting, though yielding an eventual numerical result, is still very much of a qualitative assessment. Thus, we must take all aspects of a child or parent's response into consideration. When you believe that a child's belief system concerning fire deviates considerably from the typical it should be rated C-3.

The Family Interview Form

This interview form is divided into nine content sections plus demographics. When the interview is completed, count up all C-1 responses and enter that number in the appropriate square on the Family FireRisk Summary Sheet. Repeat this process for C-2 through P-3. When complete, total each column and record that sum in the appropriate square. Once you have all totals recorded, use the total score for each of the columns to calculate the percentage of risk for child, family, and total risk according to the above formula.

The observation section of the questionnaire is filled out when you observe the family at their home. It is possible that you will choose not to interview at the home. If this is the case, skip the observation section.

It is sometimes difficult to determine when a question should receive a C-3 as opposed to a C-2 score. As an example, how long does a child have to stay and watch a fire before the behavior goes from C-2 to C-3? The answer is a function of the context. It is up to you to judge the level of dysfunction, based on your years of experience. When the length of time watching (i.e., extensive), the facial expression (i.e., transfixed), the behavior manifested (i.e., taking pictures), and general attitude suggest a "very" atypical response, you are generally warranted in giving a C-3 score.

The Parent Questionnaire

This questionnaire form is divided into eight sections. When the interview is completed, using the transparency scoring sheet, count up all C-1 responses and enter that number in the appropriate square on the Parent Questionnaire Summary Sheet. Repeat this process for C-2 through P-3. When complete, total each column and record that sum in the appropriate square. Once you have all totals recorded, use the total score for each of the columns to calculate the percentage of risk for child, family, and total risk according to the above formula.

A parent may ask for clarification on certain questions. When a parent assesses the appropriateness of a child's reaction to fire, the overall context is examined. Thus, watching the fire, running away, panicking or not, may all be C-1 responses, i.e., those responses that provide for the safety of the child as well as for others, within the child's developmental ability to provide for the safety of others. When evaluating eye contact, consider whether that behavior is appropriate to the child's culture. Severe behavior difficulties refers to extraordinary problems that a parent admits are beyond his or her ability to control. Chewing odd things has to do with those children who put things in their mouth to suck on or chew that are inappropriate considering the age of the child. Phobias refer to specific and severe fears such as heights, spiders, closed places, and snakes. General fears refers to non-specific fears.

What are excessive parental absences? A parent may ask you. This is a subjective judgment, and depends on what is normal, not so much in one family, but on what is accepted in society in general. Thus, asking whether the parent is absent from their children more than other parents in the neighborhood might be helpful.

The Structured Category Profile Sheet

At the conclusion of the interviews, transfer all individual and total scores from the Parent Questionnaire and the two evaluation forms to the Category Profile Sheet. The total scores from the summary sheets are placed in the respective subtotal columns on the structured category profile sheet. When complete, add all the columns and place the result in the total column at the bottom of the page. Next, transfer the total numeric score to compute percentages from the formula for the Child Risk, Family Risk, and Total Risk. Follow the numeric format for computing percentages from the formula. From the computation of these percentages, the child and family can be classified into risk levels.

The following criteria are used to classify the juvenile and family into risk level.

Little Risk   **Total Risk Score is equal to or less than 20%.**
Definite Risk   **Total Risk Score is between 21% - 66%.**
Extreme Risk   **Total Risk Score is equal to or greater than 67%.**

The above criteria also can be used to classify the child and family individually into their respective risk levels, however it is suggested that the Total Risk Score be used for the overall classification and recommendation for intervention and referral.

References

Fineman, K. R. (1995). A model for the qualitative analysis of child and adult fire deviant behavior. American Journal of Forensic Psychology, 13, 31-60.

Fineman, K. R. (1980). Firesetting in children and adolescents. In B. J. Blinder (Ed.), Psychiatric Clinics of North America Vol. 3. Child Psychiatry: Contributions to diagnosis, treatment, and research (pp. 483-500). Philadelphia/London/Toronto: W. B. Saunders.

COMPREHENSIVE FIRERISK EVALUATION

PARTICIPATION RELEASE

The _____utilizes the juvenile firesetter screening program developed by the Federal Emergency Management Agency and the United States Fire Administration to evaluate the child that has been involved in a fire incident or has been referred to the City by a parent or another entity or agency.

Based on the results of the evaluation, your child's tendencies will place him/her in one of the following areas of concern:

Little Risk   -   needs educational intervention

Definite Risk -   needs referral for evaluation to a mental health agency or to a licensed psychologist or psychiatrist and education intervention

Extreme Risk -   needs immediate referral for evaluation by a licensed psychologist or psychiatrist

If educational intervention is indicated, the _____ program will offer further educational activity for your child.

Depending on the circumstances regarding an individual case, other agencies such as the school your child attends, local law enforcement, social services departments, etc. may become involved.

The questions asked in this evaluation may be viewed prior to signing this release upon request.

I, _____, have read the previous statement and do hereby grant permission for my child, _____, to participate in the _____ Intervention Program and hereby authorize to release information regarding my child to such other governmental entities and agencies as it may deem appropriate.

_____   _____
     Parent/Guardian   Date/Time

_____   _____
       Juvenile   Witness

COMPREHENSIVE FAMILY FIRERISK INTERVIEW FORM
(Questions to be asked of Parents of Children 3 to 18 Years of Age)

JUVENILE FIRESETTER CONTACT FORM _____**DEPT. NAME** _____**Inc. Census Tract**_____**County**

INCIDENT-DATE_____ NO. _____ TIME_____ CR. NO_____
INCIDENT ADDRESS:_____Street _____City _____ZIP

Multiple Juveniles ☐ Y ☐ N #_____ Ignition Source: ☐ Match ☐ Lighter ☐ Other ☐ Flammable Liquid/Accelerant Used

Est. Loss: $_____ Intentional: ☐ Y ☐ N Injuries: ☐ Y ☐ N #____ Death: ☐ Y ☐ N #_____
Hospitalizations: ☐ Y ☐ N #_____ Describe Injuries/Deaths _____

Location of Fire: Outside-Location of Origin _____ ☐ Inside/☐ Inside-Occupied Room of Origin_____

Referral Source Name:_____ Agency/Address: _____ Phone: _____
   ☐ Caregiver   ☐ School   ☐ Law Enforcement   ☐ Mental Health   ☐ Fire Service   ☐ Juvenile Justice
   ☐ Parent   ☐ Other/Describe _____

Caregiver/Parent Smokes ☐ Y ☐ N Did the home meet community standards for health/welfare of the child? ☐ Y☐ N

Was the child supervised by a person 12 years of age or older at the time of the incident? ☐ Y ☐ N

Description of Incident and Pertinent Information:
_____
_____
_____

Report by: _____ _____
          Printed Name   Signature

Juvenile Information

Last Name: _____ First Name: _____ M.I. _____ DOB _____/_____/_____
Sex ☐ M ☐ F Race: ☐ White ☐ Asian ☐ African Am. ☐ Native Am. ☐ Hispanic ☐ Other
Age: _____ Grade in School_____ School Currently Attending _____
Soc. Sec. #: _____-_____-_____

Home Address: _____ Phone: _____

Adult No. 1 Residing With The Child

Name:_____

Address: _____

Phone: H _____ W _____
Employed: ☐ Y ☐ N
Marital Status: ☐ Married ☐ Separated
   ☐ Divorced ☐ Remarried ☐ Widowed

Relation to Juvenile: ☐ Natural ☐ Step

Adult No. 2 Residing With The Child

Name:_____

Address: _____

Phone: H _____ W _____
Employed: ☐ Y ☐ N
Marital Status: ☐ Married ☐ Separated
   ☐ Divorced ☐ Remarried ☐ Widowed

Relation to Juvenile: ☐ Natural ☐ Step

Others Residing With The Child

Name: _____ Relationship: _____
Name: _____ Relationship: _____
Name: _____ Relationship: _____
Name: _____ Relationship: _____

Fineman, K, (1996) *Comprehensive FireRisk Assessment.* Published in Poage, Doctor, Day, Rester, Velasquez, Moynihan, Flesher, Cooke, and Marshburn, (1997) Colorado Juvenile Firesetter Prevention Program: Training Seminar Vol 1, Denver, CO, Colorado Division of Firesafety   Comprehensive Family FireRisk Interview   Page 1 of 7

SCORE ALL ANSWERS BELOW THAT APPLY

	C-1	C-2	C-3	P-1	P-2	P-3
HEALTH HISTORY						
1. What medical or physical problems does you child have? ________ Professionally diagnosed No Yes By whom ________						
2. Has your child taken any medication in the past 3 months? If so, what? ________						
3. Has your child been diagnosed with any impulse control conditions, such as ADHD/ADD (hyperactivity)? Diagnosis ________ Yes No						
4. Is your child currently in counseling or has he/she been seen by a counselor before? For what ________ Yes (C-2) No (C-1) With whom ________						
5. Is any other family member currently in counseling or have they been seen before? By whom ________ Yes (P-2) No (P-1) For what reason ________						
6. Are there smokers in your home? Yes (P-2) No (P-1)						
Health History Subtotal						
COMMENTS:						
FAMILY STRUCTURE/ISSUES						
7. How long have you rented or owned at present location? ________ If less than 1 year score (P-2); if more that 5 years score (P-1)						
8. Do you think that you or your spouse/partner may be overprotective of the child? always (P-3) usually (P-2) sometimes (P-1) rarely (P-1) never (P-3)						
9. Is mother/female caregiver available to the child as much as the child needs her? always (P-1) usually (P-1) sometimes (P-2) rarely (P-2) never (P-3)						
10. Is father/male caregiver available to the child as much as the child needs him? always (P-1) usually (P-1) sometimes (P-2) rarely (P-2) never (P-3)						
11. Do you feel you spend enough time with your child? always (P-1) usually (P-1) sometimes (P-2) rarely (P-2) never (P-3)						
12. Are there significant conflicts between this child and other members of the family? always (P-3) usually (P-2) sometimes (P-2) rarely (P-1) never (P-1)						
13. Do you believe that you have adequate influence and control over your child? always (P-1) usually (P-1) sometimes (P-2) rarely (P-2) never (P-3)						
14. What do you discipline your child for? ________ How often? ________						
15. How do you normally discipline your child? ________						
16. Is there a history of emotional abuse the family? Yes (P-2) or (P-3) or (C-2) or (C-3) No (P-1) Who? ________ Relationship? ________ Currently in the home? ________						
17. Is there a history of physical abuse the family? Yes (P-2) or (P-3) or (C-2) or (C-3) No (P-1) Who? ________ Relationship? ________ Currently in the home? ________						
18. Is there a history of sexual abuse in the family? Yes (P-2) or (P-3) or (C-2) or (C-3) No (P-1) Who? ________ Relationship? ________ Currently in the home? ________						
Family Structure/Issues Subtotal						
COMMENTS:						

Fineman, K, (1996) *Comprehensive FireRisk Assessment.* Published in Poage, Doctor, Day, Rester, Velasquez, Moynihan, Flesher, Cooke, and Marshburn, (1997) Colorado Juvenile Firesetter Prevention Program: Training Seminar Vol 1, Denver, CO, Colorado Division of Firesafety Comprehensive Family FireRisk Interview Page 2 of 7

	C-1	C-2	C-3	P-1	P-2	P-3
PEER ISSUES						
19. Does your child interact normally with peers? Yes (C-1) No (C-2)						
20. Does your child get into fights frequently? Yes (C-2) No (C-1)						
21. Does your child frequently get picked on by other children? Yes (C-2) No (C-1)						
22. Does your child frequently play/stay alone rather than with other children? Yes (C-2) No (C-1)						
23. Do you think his friends are a bad influence? Yes (C-2) No (C-1)						
Peer Issues Subtotal						

COMMENTS:

	C-1	C-2	C-3	P-1	P-2	P-3
SCHOOL ISSUES						
24. Is your child in the age appropriate grade? Yes No If no..... [Is your child ahead (C-1) or behind (C-2)]						
25. How does your child perform academically? Well (C-1) Average (C-1) Poorly or below expectation (C-2)						
26. Have there been any recent negative changes in your child's academic performance? Yes (C-2) No (C-1)						
27. Does your child have any special educational (special ed.) learning needs? Yes [learning disabled, mentally retarded, or developmentally disabled] (C-2) No (C-1)						
28. Have there been any discipline problems at school within the last year? Yes (C-2) No (C-1)						
School Issues Subtotal						

COMMENTS:

	C-1	C-2	C-3	P-1	P-2	P-3
BEHAVIOR ISSUES						
29. Has your child been in trouble outside of school for non-fire related behaviors? What? ____ Yes (C-2) No (C-1)						
30. Does your child frequently say no when he/she is asked to do something? Yes (C-2) No (C-1)						
31. Has your child ever stolen or shoplifted? Yes (C-2) No (C-1)						
32. Has your child ever lied excessively? Yes (C-2) No (C-1)						
33. Has your child ever used drugs/alcohol/inhalants? Yes (C-2) No (C-1)						
34. Has your child ever beat up or hurt others? Yes (C-2) or (C-3) No (C-1)						
Behavior Issues Subtotal						

COMMENTS:

Fineman, K, (1996) *Comprehensive FireRisk Assessment*. Published in Poage, Doctor, Day, Rester, Velasquez, Moynihan, Flesher, Cooke, and Marshburn, (1997) Colorado Juvenile Firesetter Prevention Program: Training Seminar Vol 1, Denver, CO, Colorado Division of Firesafety Comprehensive Family FireRisk Interview Page 3 of 7

	C-1	C-2	C-3	P-1	P-2	P-3
FIRE HISTORY						
35. What were you doing when the fire occurred? appropriate supervision (P-1) not home, asleep, or other indication of inappropriate supervision, score (P-2)						
36. Are matches or lighters readily available to the child in the home? Yes (P-2) No (P-1)						
37. How did you teach your child about fire? appropriate supervision (P-1) inappropriate (P-2) e.g. has the parent directed and demonstrated proper use of fire?						
38. Have any other members of the family engaged in inappropriate fire behavior? Who? _____ Yes (P-2) No (P-1)						
39. If you had to describe your child's curiosity about fire, would you say it was: absent (C-1) mild (C-1) moderate (C-2) extreme (C-3)						
40. How many times has your child used fire inappropriately? No other times (Assess no score, skip question #41.) 1 time (C-1) 2-4 times (C-2) more than 4 times (C-3)						
Fire History Subtotal						
41. Tell me what you know about all the fires that he/she started before this one. [Use a common time frame, i.e., Christmas, school starting, etc. to help parent describe when fires were started or fireplay initiated] **INFORMATION ONLY**						

	What Set	Date Set	Where Set	With Whom	Ignition Source	Accelerant if used
1.						
2.						
3.						
4.						
5.						
Others.						

COMMENTS:

	C-1	C-2	C-3	P-1	P-2	P-3
CRISIS OR TRAUMA						
42. Has anything bad happened in the family or in your child's life within the last year? What? _____ Yes (C-2) or (P-2) No (C-1)						
43. Has there been an ongoing (chronic) crisis/problem in your child's life or in the family? Yes (C-2) or (P-2) No (C-1)						
44. Did the fire/fireplay occur after: No crisis (No score) family fight (C-2) being angry at sibling (C-2) being angry at boss (C-2) being angry at school authority (C-2) recent move (P-2) being angry with another (C-2) other crisis (C-2) or (C-3) or (P-2) or (P-3)						
Crises or Trauma Subtotal						

COMMENTS:

Fineman, K, (1996) *Comprehensive FireRisk Assessment*. Published in Poage, Doctor, Day, Rester, Velasquez, Moynihan, Flesher, Cooke, and Marshburn, (1997) Colorado Juvenile Firesetter Prevention Program: Training Seminar Vol 1, Denver, CO, Colorado Division of Firesafety   Comprehensive Family FireRisk Interview   Page 4 of 7

	C-1	C-2	C-3	P-1	P-2	P-3
CHARACTERISTICS OF FIRESTART OR FIREPLAY *[circle all that apply but only score the most severe response for each question]*						
45. Materials used to set the fire or fireplay: matches lighter flammable liquid/aerosol fireworks other (butane torch, flare, stove, pilot light) What?_____						
46. How did the child get material to start fire or engage in fireplay? found it (C-1) went out of his way to acquire it (C-2) from his hidden/saved incendiary supplies (C-2) readily available at home (P-2) or (C-1) another child had material (C-1)						
47. Where was the fire set or where did the fireplay occur? home-*occupied at the time* (C-3) other residence-*occupied at the time* (C-3) school-*occupied at the time* (C-3) other structure-*occupied at the time* (C-3) home-*unoccupied at time* (C-2) school-*unoccupied at time* (C-2) other structure-*unoccupied at time* (C-2) other residence-*unoccupied at time* (C-2) dumpster (C-2) vacant structure (C-2) outside (C-2) wildland (C-2) or (C-3) vehicle (C-2)						
48. What was set on fire? *(e.g., if the object of value was intentionally set on fire, score a C-3.)* object of little of no value (C-1) or (C-2) object of value to child (C-2) or (C-3) object of value to others (C-2) or (C-3) part of a building (C-2) people, animals, self (C-3) flammable liquids/aerosols (C-3) wildland-*unintentional* (C-2) or *intentional* (C-3) fireworks (C-2) or (P-2) paper, tissue, cardboard, twigs (C-1) or (C-2) bedding/bed-child's own (C-2) bedding/bed-someone else's (C-2) clothing-child's own (C-2) clothing - someone else's (C-2) toys (C-2) furniture (C-2) trash, leaves, grass (C-2) animals (C-3) insects (C-2) matches only (C-2) or (P-2) lighter only (C-2) or (P-2)						
49. What did he/she do after the fire started? *(If the response is appropriate based on the circumstances, score a C-1; if not, score a C-2 or C-3.)* put it out (C-1) or (C-2) called for help (C-1) ran away [if appropriate] (C-1) if not (C-2) stayed and watched (C-2) or (C-3) panicked (C-1) tried to extinguish (C-1) or (C-2) didn't try to extinguish (C-1) or (C-2) other (C-1) or (PC-2) or (C-3)						
50. Did child lie about involvement? total denial, minimizing (C-2) denial at first and then confess (C-1)						
51. Did child act alone? Yes (C-2) No (C-2) List names_____						
52. Was child pressured or coerced into firesetting or fireplay behavior by his/her peers? Yes (C-2) No (C-2) Child was instigator (C-3)						
53. Did the child respond to the fire or fireplay as if it were a positive or humorous experience? Yes (C-2) or as a negative (remorseful) experience (C-1)						
54. Does the child believe that fire has spiritual qualities or extraordinary powers? Yes (C-2) or (C-3) No (C-1)						
55. Is there an impulsive quality to the child's firesetting/fireplay? Yes (C-2) or (C-3) No (C-1)						
56. Did your child set the fire or play with fire intentionally? Yes (C-2) No (C-1)						
57. What did you do to the child in response to the fire or fireplay? grounded/restricted (P-1) physical punishment (P-1) or (P-2) nothing (P-1) or (P-2) talked/lectured (P-1) or (P-2) sought outside help (P-1) yelled (P-1) or (P-2) abused (P-2) or (P-3) other (P-1) or (P-2) Explain _____						
Characteristics of Firestart Subtotal						
COMMENTS:						

Fineman, K, (1996) *Comprehensive FireRisk Assessment.* Published in Poage, Doctor, Day, Rester, Velasquez, Moynihan, Flesher, Cooke, and Marshburn, (1997) Colorado Juvenile Firesetter Prevention Program: Training Seminar Vol 1, Denver, CO, Colorado Division of Firesafety   Comprehensive Family FireRisk Interview   Page 5 of 7

	C-1	C-2	C-3	P-1	P-2	P-3
OBSERVATIONS KEEP SEPARATE - NOT FOR PARENTAL REVIEW!						
58. How does the mother act towards the child? appropriately concerned (P-1) inappropriately concerned (P-2) hostile or indifferent (P-3)						
59. How does the father act towards the child? appropriately concerned (P-1) inappropriately concerned (P-2) hostile or indifferent (P-3)						
60. Does the mother show appropriate self-care? Yes (P-1) No (P-2)						
61. Does the father show appropriate self-care? Yes (P-1) No (P-2)						
Observations Subtotal						

COMMENTS:

Fineman, K, (1996) *Comprehensive FireRisk Assessment*. Published in Poage, Doctor, Day, Rester, Velasquez, Moynihan, Flesher, Cooke, and Marshburn, (1997) Colorado Juvenile Firesetter Prevention Program: Training Seminar Vol 1, Denver, CO, Colorado Division of Firesafety   Comprehensive Family FireRisk Interview   Page 6 of 7

COMPREHENSIVE FAMILY FIRERISK INTERVIEW SCORE SHEET

Transfer the information from the Subtotal Boxes into the table below; then total each column for the Total at the bottom.

SECTION SUBTOTALS	C-1	C-2	C-3	P-1	P-2	P-3
Health History						
Family Structure/Issues						
Peer Issues						
School Issues						
Behavior Issues						
Fire History						
Crisis or Trauma						
Characteristics of Firestart						
Observations						
TOTAL						

These totals will be used to compute the Total Risk after all interviews are complete.

Fineman, K, (1996) *Comprehensive FireRisk Assessment.* Published in Poage, Doctor, Day, Rester, Velasquez, Moynihan, Flesher, Cooke, and Marshburn, (1997) Colorado Juvenile Firesetter Prevention Program: Training Seminar Vol 1, Denver, CO, Colorado Division of Firesafety   Comprehensive Family FireRisk Interview   Page 7 of 7

COMPREHENSIVE CHILD FIRERISK INTERVIEW FORM
(Questions to be asked of Children 3 to 18 Years of Age)

AGENCY _____   COUNTY _____

INTERVIEWER _____ DATE _____

JUVENILE'S NAME _____

SEX _____ DOB _____ ETHNICITY/RACE _____

ADDRESS _____ PHONE _____

SCHOOL _____ GRADE _____

DEVELOPMENT OF RAPPORT

The purpose of this section is to make the child comfortable with you. The more at ease you can make him, the greater the likelihood that he will answer all of your questions. If the following questions aren't enough, add your own. Questions or language can be modified throughout this form to accommodate the age of the child or adolescent.

A. [Introduce yourself] I'm _____ What's your name?_____

B. How old are you? _____

C. What school do you go to? _____ What grade are you in? _____

D. Do you like your school?_____ Are there nice/okay teachers at your school?____

E. What classes/subjects do you like/not like? _____

F. What do you do for fun? Do you have hobbies? _____

G. Who's you best friend? _____

H. What do you like to play/do with your friend? _____

I. What do you watch on TV and/or what videos do you watch? _____

J. What is your favorite person/show on TV? _____

K. What is your favorite video/computer game? _____

L. What do you like about that game? [Is there extreme interest in violence or fire?] _____

[When rapport is established, determine level of understanding if the child is under 7 or appears to have problems communicating.]

Fineman, K, (1996) *Comprehensive FireRisk Assessment.* Published in Poage, Doctor, Day, Rester, Velasquez, Moynihan, Flesher, Cooke, and Marshburn, (1997) Colorado Juvenile Firesetter Prevention Program: Training Seminar Vol 1, Denver, CO, Colorado Division of Firesafety   Comprehensive Child FireRisk Interview   Page 1 of 7

DETERMINE LEVEL OF UNDERSTANDING

It is often difficult to determine if a young child really understands you. (These instructions may be skipped if you are interviewing an older child.) There may be an age barrier, a language barrier, a learning problem, or sub-normal intelligence. It is fruitless to go through an entire interview unless you are first assured that the child has enough understanding to complete the interview. There are several ways to gauge whether you are on the same "wave length" as the child. The following are suggested ways to do so:

a. Obtain information from rapport section above:
     By paying close attention to the manner in which a young child responds to the 11 questions above, you can estimate whether he can understand and respond to the other questions in this instrument.

b. Using crayons/paper as a tool:
     You can ask the child to draw pictures of common objects, his favorite toys, houses, trees, and people. Then ask him to describe what he has drawn. Clear explanations of his drawings and the action taking place in some of those drawings will tell you something about the child's vocabulary and his ability to understand.

c. Using toys and games:
     Have toys of the appropriate developmental level of the child available. Engage the child in a game with the toys or allow the child free play with the toys. After a while ask the child about the toys and the game he/she is playing. Inquire about the rules, the purpose, etc. Estimate the child's vocabulary in terms of his/her ability to complete the interview.

d. Using puppets:
     Have hand puppets available. Allow the child to set the interaction, with the child playing all parts or with you playing some of the parts. Quiet children can become quite verbal with this approach. Focus on the child's ability to understand your questions during the puppet play and determine if this level of communication is sufficient for continued interviewing.

If you are satisfied that the child has adequate understanding, proceed with the interview.

SCORE ALL ANSWERS BELOW THAT APPLY

		C-1	C-2	C-3	P-1	P-2	P-3
SCHOOL ISSUES (If home schooled, skip question #2)							
1. Do you like school/learning? Yes (C-1) No (C-2)							
2. Do you listen to your teacher(s) most of the time: Yes (C-1) No (C-2)							
3. Have there been any recent problems with your school performance within the last year? Yes (C-2) No (C-1)							
4. Have you gotten in trouble at school? Yes (C-2) No (C-1)							
School Issues Subtotal							
COMMENTS:							
PEER ISSUES							
5. Do you get along with most of your friends? Yes (C-1) No (C-2)							
6. Do you get picked on? Yes (C-2) No (C-1)							
7. Do you have as many friends as you want? Yes (C-1) No (C-2)							
8. Do you want to be alone or with other kids? Alone (C-2) Kids (C-1)							
9. Do you think your friends are a bad influence on you? Yes (C-2) No (C-1)							
Peer Issues Subtotal							
COMMENTS:							

Fineman, K, (1996) *Comprehensive FireRisk Assessment.* Published in Poage, Doctor, Day, Rester, Velasquez, Moynihan, Flesher, Cooke, and Marshburn, (1997) Colorado Juvenile Firesetter Prevention Program: Training Seminar Vol 1, Denver, CO, Colorado Division of Firesafety   Comprehensive Child FireRisk Interview   Page 2 of 7

	C-1	C-2	C-3	P-1	P-2	P-3
BEHAVIOR ISSUES						
10. Do you get in trouble frequently at school? Yes (C-2) No (C-1)						
11. Do you usually not do things that you are asked to do? Yes (C-2) No (C1)						
12. Have you ever stolen or shoplifted? Yes (C-2) No C-1)						
13. Have you ever frequently lied? Yes (C-2) No (C-1)						
14. Have you ever used drugs, alcohol, or inhalants? Yes (C-2) No (C-1)						
15. Have you ever beat up or hurt others? Yes (C-2) or (C-3) No (C-1)						
Behavior Issues Subtotal						
COMMENTS:						
FAMILY ISSUES						
16. Do you like going home? Yes No Why?_____						
17. How well do you get along with your mother (female caregiver)? always get along (P-1) usually get along (P-1) sometimes get along (P-2) don't get along very often (P-2) never get along (P-3)						
18. Do you fight or argue with your mother? always (P-3) usually (P-2) sometimes (P-1) rarely (P-1) never (P-1)						
19. Are you afraid of your mother? always (P-3) usually (P-2) sometimes (P-2) rarely (P-1) never (P-1)						
20. How well do you get along with your father (male caregiver)? always get along (P-1) usually get along (P-1) sometimes get along (P-2) don't get along very often (P-2) never get along (P-3)						
21. Do you fight or argue with your father? always (P-3) usually (P-2) sometimes (P-1) rarely (P-1) never (P-1)						
22. Are you afraid of your father? always (P-3) usually (P-2) sometimes (P-2) rarely (P-1) never (P-1)						
23. Do your mother and father fight? [If the parents fight, have the child elaborate on the fights] always (P-3) usually (P-2) sometimes (P-1) rarely (P-1) never (P-1)						
24. Tell me abut your brothers and/or sisters. How well do you get along with them? *(If there is a variability in the relationship among siblings, rate the most serious.)* always get along (P-1) usually get along (P-1) sometimes get along (P-2) don't get along very often (P-2) never get along (P-3)						
25. Do you see your mom as much as you'd like? Yes (P-1) No (P-2)						
26. Do you see your dad as much as you'd like? Yes (P-1) No (P-2)						
27. What do you do that gets you into trouble at home? _____						
28. What happens at home when you get in trouble? grounded him/her (P-1) physical punishment (P-1) or (P-2) nothing (P-2) talked/lectured (P-1) or (P-2) sought outside help (P-1) yelled (P-1) or (P-2) abused (P-2) or (P-3) other (P-1) or (P-2) Explain_____						
29. Do you get spanked/punished too much? Yes (P-2) No (P-1) If so, by whom_____						
Family Issues Subtotals						
COMMENTS:						

Fineman, K, (1996) *Comprehensive FireRisk Assessment*. Published in Poage, Doctor, Day, Rester, Velasquez, Moynihan, Flesher, Cooke, and Marshburn, (1997) Colorado Juvenile Firesetter Prevention Program: Training Seminar Vol 1, Denver, CO, Colorado Division of Firesafety Comprehensive Child FireRisk Interview Page 3 of 7

	C-1	C-2	C-3	P-1	P-2	P-3
CRISIS OR TRAUMA (Probe for severity)						
30. Within the last year has anything bad happened in your life? Yes (C-2) or (P-2)　　No (C-1)　　What?_____						
31. Has there been an ongoing (chronic) crisis/problem in your life or in the family? Yes (C-2) or (P-2)　　No (C-1)　　What?_____						
32. Was the fire set after:　　No crisis (no score)　　family fight (C-2) being angry at sibling (C-2)　　being angry with boss (C-2)　　being angry with school authority (C-2)　　being angry with another (C-2)　　recent move (P-2) other crises, such as stress, death, depression (C-2) or (C-3) or (P-2) or (P-3) What?_____						
Crisis or Trauma Subtotal						

COMMENTS:

	C-1	C-2	C-3	P-1	P-2	P-3
FIRE HISTORY						
33. Do you like to look at fire for long periods of time?　　Yes (C-2) or (C-3)　　No (C-1)						
34. Do you dream about fires at night?　　Yes (C-2) or (C-3)　　No (C-1)						
35. Do you think about or daydream about fires in the day?　Yes (C-2) or (C-3)　　No (C-1)						
36. Number of past (inappropriate) fires or fireplay incidents　No other times (Assess no score, skip question #37.)　　1 time (C-1)　　2-4 times (C-2)　　more than 4 times (C-3)						
37. Tell me about all the fires that you started or your fireplay before this one. [Use a common time frame, i.e., Christmas, school starting, etc. to help child describe when fires were started or fireplay occurred]　　**INFORMATION ONLY**						

What Set	Date Set	Where Set	With Whom	Ignition Source	Accelerant if used
1.					
2.					
3.					
4.					
5.					
Others.					

	C-1	C-2	C-3	P-1	P-2	P-3
38. *If there is more than one fire ask questions #38 and #39.* Do you feel the need to set fires over and over again?　　Yes (C-2) or (C-3)　　No (C-1)						
39. Do you always set your fires in exactly the same way?　　Yes (C-2) or (C-3)　　No (C-1)						
Fire History Subtotals						

COMMENTS:

Fineman, K. (1996) *Comprehensive FireRisk Assessment.* Published in Poage, Doctor, Day, Rester, Velasquez, Moynihan, Flesher, Cooke, and Marshburn, (1997) Colorado Juvenile Firesetter Prevention Program: Training Seminar Vol 1, Denver, CO, Colorado Division of Firesafety　　Comprehensive Child FireRisk Interview　　Page 4 of 7

	C-1	C-2	C-3	P-1	P-2	P-3
CHARACTERISTICS OF FIRESTART OR FIREPLAY [circle all that apply but only score the most severe response for each question]						
40. Tell me about how you think the fire/fireplay started? admits/confesses (C-1) denies or minimizes (C-2) denial then truth (C-1)						
41. What do you think made you want to start the fire or the fireplay/what happened? to express anger (C-2) or (P-2) to see it burn (C-2) bored (C-2) to show power or control (C-2) or (C-3) didn't want to (accident or curiosity) (C-1) reaction to stress (C-2) or (P-2) from peer pressure (C-2) to destroy something (C-2) or (C-3) or (P-2) to hurt self (C-3) or (P-2) to hurt others (C-2) or (P-2) to get attention (C-2) or (P-2) don't know (C-2) rebellion - was told not to do so (C-2) or (P-2)						
42. What did you use to set the fire or start the fireplay? matches lighter flammable liquid/aerosol fireworks flarestove pilot light other_____						
43. How did you get the (above igniter) to start the fire or the fireplay? went out of way to acquire (C-2) found it (C-1) hidden stockpile (C-2) readily available at home (P-2) or (C-1) another child had material (C-1)						
44. What was set on fire? *(e.g., if the object of value was incidental to the fire score a C-2; or if purposely set on fire score a C-3.)* object of little or no value (C-1) or (C-2) object of value to child (C-2) or (C-3) object of value to others (C-2) or (C-3) part of a building (C-2) people, animals, self (C-3) flammable liquids/aerosols (C-3) wildland *intentional* (C-2) or *intentional* (C-3) fireworks (C-2) or (P-2) paper, tissue, cardboard, twigs (C-1) or (C-2) bedding/bed-child's own (C-2) bedding/bed-someone else's (C-2) clothing-child's own (C-2) clothing-someone else's (C-2) toys (C-2) furniture (C-2) trash, leaves, grass (C-2) animals (C-3) insects (C-2) matches only (C-2) or (P-2) lighter only (C-2) or (P-2)						
45. Where was the fire set or where did the fireplay occur? home-*occupied at the time* (C-3) other residence-*occupied at the time* (C-3) school-*occupied at the time* (C-3) other structure-*occupied at the time* (C-3) home-*unoccupied at time* (C-2) school-*unoccupied at time* (C-2) other structure-*unoccupied at time* (C-2) other residence-*unoccupied at time* (C-2) dumpster (C-2) vacant structure (C-2) outside (C-2) wildland (C-2) or (C-3) vehicle (C-2)						
46. Did you intend to set the fire? Yes (C-2) No (C-1)						
47. Did you drink or take any drugs before, during, or after the fire/fireplay Yes (C-2) No (C-1)						
48. What did you do after the fire started? *(If the response is appropriate based on the circumstances, score a C-1; if not, score a C-2 or C-3.)* put it out (C-1) or (C-2) called for help (C-1) ran away [if appropriate] (C-1) if not (C-2) stayed and watched (C-2) or (C-3) panicked (C-1) tried to extinguish (C-1) or (C-2) didn't try to extinguish (C-1) or (C-2) other (C-1) or (PC-2) or (C-3)						
49. How do your parents punish you? grounded/restricted (P-1) physical punishment (P-1) or (P-2) nothing (P-1) or (P-2) talked/lectured (P-1) or (P-2) sought outside help (P-1) yelled (P-1) or (P-2) abused (P-2) or (P-3) other (P-1) or (P-2) Explain _____						

Fineman, K, (1996) *Comprehensive FireRisk Assessment.* Published in Poage, Doctor, Day Rester, Velasquez, Moynihan, Flesher, Cooke, and Marshburn, (1997) Colorado Juvenile Firesetter Prevention Program: Training Seminar Vol 1, Denver, CO, Colorado Division of Firesafety Comprehensive Child FireRisk Interview Page 5 of 7

	C-1	C-2	C-3	P-1	P-2	P-3
50. Did the fire(s) or fireplay you started make you happy or make you laugh? Yes (C-3) No (C-1)						
51. Can fire do magical, special, or miraculous things? Yes (C-2) or (C-3) No (C-1) Explain _____						
52. After the fire how did you feel? happy (C-2) nervous (C-1) sad (C-1) powerful (C-3) angry (C-2) hateful (C-2) vengeful (C-2) scared (C-1) remorseful (C-1) elated (C-3) guilty (C-1) ashamed (C-1) excited (C-3) curious (C-1) or (C-3) aroused sexually (C-3) aroused sensually (C-3)						
Characteristics of Firestart Subtotal						
COMMENTS:						
OBSERVATIONS KEEP SEPARATE - NOT FOR PARENTAL REVIEW!						
53. Are child's behaviors and mannerisms: normal (C-1) troubled (C-2) very troubled (C-3)						
54. Is the child's mood: normal (C-1) troubled (C-2) very troubled (C-3)						
55. Is the child's way of thinking: normal (C-1) troubled (C-2) very troubled (C-3)						
56. Are there signs of abuse? Yes (P-2) or (P-3) No (P-1) Explain_____						
57. Are there signs of neglect? Yes (P-2) or (P-3) No (P-1) Explain_____						
Observations Subtotal						
COMMENTS:						

Fineman, K, (1996) *Comprehensive FireRisk Assessment*. Published in Poage, Doctor, Day Rester, Velasquez, Moynihan, Flesher, Cooke, and Marshburn, (1997) Colorado Juvenile Firesetter Prevention Program: Training Seminar Vol 1, Denver, CO, Colorado Division of Firesafety   Comprehensive Child FireRisk Interview   Page 6 of 7

Comprehensive Juvenile FireRisk Interview Form Score Sheet

Transfer the information from the Subtotal Boxes into the table below; then total each column for the Total at the bottom.

SECTION SUBTOTALS	C-1	C-2	C-3	P-1	P-2	P-3
School Issues						
Peer Issues						
Behavior Issues						
Family Issues						
Crisis or Trauma						
Fire History						
Characteristics of Firestart						
Observations						
TOTAL						

These totals will be used to compute the Total Risk after all interviews are complete.

Fineman, K , (1996) *Comprehensive FireRisk Assessment.* Published in Poage, Doctor, Day, Rester, Velasquez, Moynihan, Flesher, Cooke, and Marshburn, (1997) Colorado Juvenile Firesetter Prevention Program: Training Seminar Vol I , Denver, CO, Colorado Division of Firesafety   Comprehensive Child FireRisk Interview   Page 7 of 7

COMPREHENSIVE PARENT FIRERISK QUESTIONNAIRE
for the child 3 to 18 years of age

Respondent _____ Agency _____ County _____ Date _____

PARENTS: Please complete this form. Mark the answer under "rarely to never," "sometimes," or "frequently" that best describes your child for each question. When marking the form, consider all parts of the child's life (at home, at school, etc.) where the events below might occur. If an item does not apply, leave it blank. If you do not understand a term or question, make a mark next to it in the left margin and ask the interviewer for clarification.

ITEM	RARELY TO NEVER	SOMETIMES	FREQUENTLY
Hyperactivity at school			
Lack of concentration			
Learning problems at school			
Behavior problems at school			
Impulsive (acts before he thinks)			
Impatient			
Fantasizes (daydreaming)			
Likes school			
Listens to teacher(s)/school authorities			
Shows age appropriate interest in future school/jobs/career			
Truant/school runaway			
Convulsions, seizures, "spells"			
Need for excessive security			
Need for affection			
Loss of appetite			
Excessive weight loss			
Excessively overweight			
Knows what is moral			
Feels good about self			
Comfortable with own body			
Likes overall looks			
Stuttering			
Wets during the day (after age 3)			
Night time bed wetting (after age 3)			
Soiling (after age 3)			
Participates in sports			

Fineman, K , (1996) *Comprehensive FireRisk Assessment.* Published in Poage, Doctor, Day, Rester, Velasquez, Moynihan, Flesher, Cooke, and Marshburn, (1997) Colorado Juvenile Firesetter Prevention Program: Training Seminar Vol I , Denver, CO, Colorado Division of Firesafety Comprehensive Parent FireRisk Questionnaire Page 1 of 5

ITEM	RARELY TO NEVER	SOMETIMES	FREQUENTLY
Injury prone			
Shyness			
Tries to please everyone			
Relationships are socially appropriate			
Physically fights with peers			
Withdraws from peers/group			
Destroys toys/property of others			
A poor loser			
Shows off for peers			
Easily led by peers			
Plays with other children			
Shows appropriate peer affection			
Plays alone (not even with adults)			
Picked on by peers			
Has many friends			
Is good at sports			
Is a loner (few friends)			
Lies			
Excessive and uncontrolled verbal anger			
Physically violent			
Steals			
Cruel to animals			
Cruel to children			
Is/was in a gang			
Expresses anger by damaging the property of others			
Destroys own toys/possessions (if child is age 3-6)			
Destroys own toys/possessions (if child is age 7-18)			
Disobeys			
Severe behavior difficulties (past or present)			
Expresses anger by hurting others' things			
Has been in trouble with police			

Fineman, K , (1996) *Comprehensive FireRisk Assessment.* Published in Poage, Doctor, Day, Rester, Velasquez, Moynihan, Flesher, Cooke, and Marshburn, (1997) Colorado Juvenile Firesetter Prevention Program: Training Seminar Vol I , Denver, CO, Colorado Division of Firesafety   Comprehensive Parent FireRisk Questionnaire   Page 2 of 5

	RARELY TO NEVER	SOMETIMES	FREQUENTLY
Uses drugs or alcohol			
Jealous of peers/siblings			
Temper tantrums			
Unacceptable showing off			
Sexual activity with others			
Stomach aches			
Nightmares			
Sleeps too deep or has problem waking up			
Anxiety (nervousness)			
Has twitches (eyes, face, etc.)			
Cries			
Bites nails			
Vomits			
Aches and pains			
Chews odd/unusual things			
Extreme mood swings			
Depressed mood or withdrawal			
Constipation			
Diarrhea			
Self-imposed unnecessary or excessive diets			
Sleepwalking			
Phobias			
General fears			
Curiosity about fire			
Plays with matches/lighters			
Plays with fire (singeing, burning)			
Was concerned when fire got out of control			
Was proud or boastful regarding fireplay or firestart			
Stares at fire for long periods (fire fascination)			
Unusual look on child's face when he/she stares at fire(s)			
Daydreams or talks about fires			

Fineman, K , (1996) *Comprehensive FireRisk Assessment.* Published in Poage, Doctor, Day, Rester, Velasquez, Moynihan, Flesher, Cooke, and Marshburn, (1997)
Colorado Juvenile Firesetter Prevention Program: Training Seminar Vol I , Denver, CO, Colorado Division of Firesafety   Comprehensive Parent FireRisk
Questionnaire   Page 3 of 5

ITEM	RARELY TO NEVER	SOMETIMES	FREQUENTLY
Fear of fire			
Other(s) in family set fire(s) (past or present)			
Set occupied structure on fire			
Appropriate reaction to fire(s) he/she set			
Extensive absences by father			
Extensive absences by mother			
Family has moved			
Runs away from home			
Has seen a counselor/therapist			
Other family member has seen a counselor/therapist			
Makes attempts at age appropriate independence from parents			
In trouble at home			
Parent or sibling with serious health problem			
Marriage is unhappy			
Mother's discipline is effective			
Father's discipline is effective			
Fighting with siblings			
Conflicts in family			
Unusual fantasies			
Strange thought patterns			
Bizarre, illogical, or irrational speech			
Out of touch with reality			
Strange quality about child			
Expresses anger by hurting self or something he/she likes			
Destroys own property			
Was/is in a cult			
Severe depression or withdrawal			
Poor or no eye contact			

Fineman, K , (1996) *Comprehensive FireRisk Assessment.* Published in Poage, Doctor, Day, Rester, Velasquez, Moynihan, Flesher, Cooke, and Marshburn, (1997) Colorado Juvenile Firesetter Prevention Program: Training Seminar Vol I , Denver, CO, Colorado Division of Firesafety   Comprehensive Parent FireRisk Questionnaire   Page 4 of 5

PARENT QUESTIONNAIRE SCORE SHEET

Transfer the information you obtained above to the table below; then total each column for the Total at the bottom.

	C-1	C-2	C-3	P-1	P-2	P-3
School						
Health/Developmental						
Peers						
Antisocial Behavior (BEHAVIOR)						
Symptoms of Anxiety or Depression (ANXIETY)						
Fire History						
Family Issues (FAMILY)						
Severe Dysfunction (OTHER)						
TOTAL						

These totals will be used to compute the Total Risk after all interviews are complete.

Fineman, K , (1996) *Comprehensive FireRisk Assessment.* Published in Poage, Doctor, Day, Rester, Velasquez, Moynihan, Flesher, Cooke, and Marshburn, (1997) Colorado Juvenile Firesetter Prevention Program: Training Seminar Vol I , Denver, CO, Colorado Division of Firesafety   Comprehensive Parent FireRisk Questionnaire   Page 5 of 5

COMPREHENSIVE PARENT FIRERISK QUESTIONNAIRE
for the child 3 to 18 years of age

VISUAL KEY

	RARELY TO NEVER	SOMETIMES	FREQUENTLY
SCHOOL			
Hyperactivity at school			C-2
Lack of concentration	C-1	C-1	C-2
Learning problems at school		C-2	C-2
Behavior problems at school	C-1	C-2	C-2
Impulsive (acts before he thinks)	C-1	C-1	C-2
Impatient	C-1	C-1	C-2
Fantasizes (daydreaming)			C-2
Likes school	C-2	C-1	C-1
Listens to teacher(s)/school authorities	C-2		C-1
Shows age appropriate interest in future school/jobs/career	C-2	C-1	C-1
Truant/school runaway		C-2	C-3
HEALTH/DEVELOPMENTAL			
Convulsions, seizures, "spells"		C-2	C-2
Need for excessive security	C-2	C-1	C-2
Need for affection	C-2	C-1	C-2
Loss of appetite			C-2
Excessive weight loss		C-2	C-2
Excessively overweight			C-2
Knows what is moral	C-2		C-1
Feels good about self	C-2		C-1
Comfortable with own body	C-2		C-1
Likes overall looks	C-2		C-1
Stuttering		C-2	C-2
Wets during the day (after age 3)	C-1	C-2	C-2
Night time bed wetting (after age 3)	C-1	C-2	C-2
Soiling (after age 3)		C-2	C-2
Participates in sports	C-2		C-1

Fineman, K , (1996) *Comprehensive FireRisk Assessment.* Published in Poage, Doctor, Day, Rester, Velasquez, Moynihan, Flesher, Cooke, and Marshburn, (1997) Colorado Juvenile Firesetter Prevention Program: Training Seminar Vol I , Denver, CO, Colorado Division of Firesafety   Comprehensive Parent FireRisk Questionnaire Key   Page 1 of 4

	RARELY TO NEVER	SOMETIMES	FREQUENTLY
Injury prone	C-1		C-2
Shyness	C-1		C-2
Tries to please everyone			C-2
Relationships are socially appropriate	C-2		C-1
PEERS			
Physically fights with peers	C-1		C-2
Withdraws from peers/group	C-1		C-2
Destroys toys/property of others	C-1	C-2	C-2
A poor loser	C-1		C-2
Shows off for peers			C-2
Easily led by peers	C-1	C-2	C-3
Plays with other children	C-2		C-1
Shows appropriate peer affection	C-2		C-1
Plays alone (not even with adults)	C-1		C-2
Picked on by peers	C-1		C-2
Has many friends	C-2	C-1	C-1
Is good at sports	C-2		C-1
Is a loner (few friends)	C-1	C-2	C-3
BEHAVIOR			
Lies	C-1		C-2
Excessive and uncontrolled verbal anger	C-1	C-2	C-3
Physically violent	C-1	C-2	C-3
Steals	C-1	C-2	C-3
Cruel to animals		C-2	C-3
Cruel to children		C-2	C-3
Is/was in a gang		C-2	C-3
Expresses anger by damaging the property of others			C-2
Destroys own toys/possessions (if child is age 3-6)			C-2
Destroys own toys/possessions (if child is age 7-18)		C-2	C-3
Disobeys	C-1		C-2
Severe behavior difficulties (past or present)		C-2	C-3
Expresses anger by hurting others' things		C-2	C-3
Has been in trouble with police		C-2	C-3

Fineman, K., (1996) *Comprehensive FireRisk Assessment.* Published in Poage, Doctor, Day, Rester, Velasquez, Moynihan, Flesher, Cooke, and Marshburn, (1997) Colorado Juvenile Firesetter Prevention Program: Training Seminar Vol 1, Denver, CO, Colorado Division of Firesafety   Comprehensive Parent FireRisk Questionnaire Key   Page 2 of 4

	RARELY TO NEVER	SOMETIMES	FREQUENTLY
Uses drugs or alcohol		C-2	C-3
Jealous of peers/siblings	C-1		C-2
Temper tantrums	C-1		C-2
Unacceptable showing off	C-1		C-2
Sexual activity with others		C-3	C-3
ANXIETY			
Stomach aches			C-2
Nightmares	C-1		C-2
Sleeps too deep or has problem waking up		C-2	C-2
Anxiety (nervousness)	C-1		C-2
Has twitches (eyes, face, etc.)		C-2	C-2
Cries			C-2
Bites nails			C-2
Vomits			C-2
Aches and pains			C-2
Chews odd/unusual things			C-2
Extreme mood swings		C-2	C-2
Depressed mood or withdrawal		C-2	C-3
Constipation			C-2
Diarrhea			C-2
Self-imposed unnecessary or excessive diets			C-2
Sleepwalking		C-2	C-2
Phobias		C-2	C-3
General fears	C-1		C-2
FIRE HISTORY			
Curiosity about fire	C-1		C-2
Plays with matches/lighters	C-1	C-2	C-3
Plays with fire (singeing, burning)	C-1	C-2	C-3
Was concerned when fire got out of control	C-3	C-2	C-1
Was proud or boastful regarding fireplay or firestart		C-3	C-3
Stares at fire for long periods (fire fascination)		C-2	C-3
Unusual look on child's face when he/she stares at fire(s)		C-2	C-3
Daydreams or talks about fires		C-2	C-3

Fineman, K , (1996) *Comprehensive FireRisk Assessment.* Published in Poage, Doctor, Day, Rester, Velasquez, Moynihan, Flesher, Cooke, and Marshburn, (1997)
Colorado Juvenile Firesetter Prevention Program: Training Seminar Vol I , Denver, CO, Colorado Division of Firesafety   Comprehensive Parent FireRisk
Questionnaire Key   Page 3 of 4

	RARELY TO NEVER	SOMETIMES	FREQUENTLY
Fear of fire	C-2		C-1
Other(s) in family set fire(s) (past or present)		P-2	P-3
Set occupied structure on fire		C-3	C-3
Appropriate reaction to fire(s) he/she set	C-3	C-2	C-1
FAMILY			
Extensive absences by father	P-1	P-2	P-2
Extensive absences by mother	P-1	P-2	P-2
Family has moved			P-2
Runs away from home	C-1	C-2	C-2
Has seen a counselor/therapist		C-2	C-2
Other family member has seen a counselor/ therapist		P-2	P-2
Makes attempts at age appropriate independence from parents	C-2	C-1	C-1
In trouble at home	C-1		C-2
Parent or sibling with serious health problem		P-2	P-2
Marriage is unhappy	P-1		P-2
Mother's discipline is effective	P-2		P-1
Father's discipline is effective	P-2		P-1
Fighting with siblings	C-1		C-2
Conflicts in family	P-1		P-2
OTHER			
Unusual fantasies		C-2	C-3
Strange thought patterns		C-2	C-3
Bizarre, illogical, or irrational speech		C-3	C-3
Out of touch with reality		C-3	C-3
Strange quality about child		C-2	C-3
Expresses anger by hurting self or something he/she likes		C-3	C-3
Destroys own property			C-2
Was/is in a cult		C-2	C-3
Severe depression or withdrawal		C-3	C-3
Poor or no eye contact		C-2	C-2

Fineman, K , (1996) *Comprehensive FireRisk Assessment.* Published in Poage, Doctor, Day, Rester, Velasquez, Moynihan, Flesher, Cooke, and Marshburn, (1997) Colorado Juvenile Firesetter Prevention Program: Training Seminar Vol I , Denver, CO, Colorado Division of Firesafety   Comprehensive Parent FireRisk
Questionnaire Key   Page 4 of 4

THE STRUCTURED CATEGORY PROFILE SHEET

COMPREHENSIVE FIRERISK ANALYSIS

Transfer the values from the "TOTAL" line for the family interview, parent questionnaire, and the child interview to the table below; add the columns for a "GRAND TOTAL." Use these totals to compute the percentages according to the formula below the table.

	C-1	C-2	C-3	P-1	P-2	P-3
Family Interview TOTAL						
Parent Questionnaire TOTAL						
Child Interview TOTAL						
GRAND TOTAL						

Child Risk (Use the values from the Grand Total Line)

$$\frac{C-2+C-3}{C-1+C-2+C-3} = \underline{\hspace{2cm}}\%$$

Family Risk (Use the values from the Grand Total Line.)

$$\frac{P-2+P-3}{P-1+P-2+P-3} = \underline{\hspace{2cm}}\%$$

Total Risk (Use the values from the Grand Total Line.)

$$\frac{C-2+P-2+C-3+P-3}{C-1+P-1+C-2+P-2+C-3+P-3} = \underline{\hspace{2cm}}\%$$

Fineman, K , (1996) *Comprehensive Fire Risk Assessment.* Published in Poage, Doctor, Day, Rester, Velasquez, Moynihan, Flesher, Cooke, and Marshburn, (1997) Colorado Juvenile Firesetter Prevention Program: Training Seminar Vol 1 Denver, CO, Colorado Division of Firesafety   Comprehensive FireRisk Analysis
Page 1 of 1

ENCUESTA DEL PADRE
por el niño 3-18

Respondent _______________ **Agency** _______________ **County** _________ **Date** _____

PADRES: Favor de completar esta forma. Mark la repuesta debajo de "raramente a nunca," "a veces" o "frecuentemente" ese major describe a su niño por casa pregunta. Cuando marca la forma, considera todo partes de la vida de la niño (en casa, a escuela, etc.) donde los eventos abajo pueden ocurrir. Si un artículo no aplica, sale borra. Si no entiende un término o pregunta, hechura una marca al lado de él en el margen izquierdo y pregunta el entrevistador por clarificación.

ARTICULO	RARAMENTE A NUNCA	A VECES	FRECUENTE-MENTE
Hyperactividad a escuela			
Falte de concentración			
Problemas del aprender a escuela			
Problemas de la conducta en escuela			
Impulsivo (actos antes de piensa)			
Impaciente			
Fantasizes (día soña)			
Le gusta la escuela			
Escucha a maestro(s)/autoridades escolares			
Le muestra interés apropiado en futuro a edad escuela/trabajos/carrera			
Truant/fugitivo de la escuela			
Convulsiones, cogidos, "hechizos"			
Necesidad por garantía excesiva			
Necesidad por afecto			
Pérdida de apetito			
Pérdida del peso excesiva			
Sobrepeso excesivo			
Sabe qué es moral			
Tactos bueno acerca de mismo			
Cómodo con cuerpo propio			
Gusta miradas del todo			
Tartamudear			
Moja durante el día (después de edad 3)			
Cama del tiempo de la noche moja (después de edad 3)			
Ensuciar (después de edad 3)			
Participa en deportes			

ARTICULO	RARAMENTE A NUNCA	A VECES	FRECUENTE- MENTE
Lesión prono			
Timidez			
Trata de agradar a toda la gente			
Relaciones son apropiado socialmente			
Lucha con pares (fisico)			
Retira de pares/grupo			
Destruye juguetes/propiedad de otros			
Un perdedor pobre			
Muestra apartado por pares			
Fácilmente llevó por pares			
Juega a con otro niños			
Muestra afecto del par apropiado			
Juga a solo (tampoco con adultos)			
Recogió en por pares			
Tiene muchos amigos			
Está hábil en deportes			
Es un solitario (pocos amigos)			
Mentiras			
Excesivo y enojo libre verbal			
Fisico violento			
Robos			
Cruel a animales			
Cruel a niños			
Está/estaba en una pelotón			
Enojo de los expresos por prejudicial en propiedad de otros			
Destruye juguetes propios/posesiones (si niño está edad 3-6)			
Destruye juguetes propios/posesiones (si niño está edad 7-18)			
Desobedece			
Dificultades de la conducta severas (pasado o presente)			
Enojo do los expresos por herir cosas del otros			
Ha estado en problema con policiá			

Fineman, K , (1996) *Comprehensive Fire Risk Assessment.* Published in Poage, Doctor, Day, Rester, Velasquez, Moynihan, Flesher, Cooke, and Marshburn, (1997) Colorado Juvenile Firesetter Prevention Program: Training Seminar Vol 1 Denver, CO, Colorado Division of Firesafety   Parent FireRisk Questionnaire
Page 2 of 5

ARTICULO	RARAMENTE A NUNCA	A VECES	FRECUENTE-MENTE
Usa drogas o alcohol			
Celoso de pares/hermanos			
Rabietas del temple			
Exhibición inaceptable			
Actividad sexual con otros			
Dolores del estómago			
Pesadillas			
Duerme demasiado hondo o problema se despierta			
Ansiedad (nerviosidad)			
Tiene tirones (ojoes, cara, etc.)			
Lamentos			
Uñas de las mordeduras			
Vómitos			
Dolores y dolores			
Mastica impar/cosas raras			
Humor extremo gira			
Deprimió humor o retiro			
Estreñimiento			
Diarrea			
Mismo-impuso dietas innecesarias o excesivas			
Sonambulo			
Fobias			
Miedos generales			
Curiosidad acerca de fuego			
Juego a con fósforos/encendedores			
Juega con fuego (chamusco, ardiente)			
Se concernió cuando fuego hizo fuera de mando			
Estaba orgulloso o jactancioso con respecto del fuego			
Mirada fija a fuego por periodos largos (fascinación del fuego)			
Raro parece en niño cara cuando él/ella mirada fija a fuego			
Ensueños o habla acerca de fuegos			

Fineman, K , (1996) *Comprehensive Fire Risk Assessment.* Published in Poage, Doctor, Day, Rester, Velasquez, Moynihan, Flesher, Cooke, and Marshburn, (1997) Colorado Juvenile Firesetter Prevention Program: Training Seminar Vol 1 Denver, CO, Colorado Division of Firesafety   Parent FireRisk Questionnaire Page 3 of 5

ARTICULO	RARAMENTE A NUNCA	A VECES	FRECUENTE-MENTE
Miedo de fuego			
Otro(s) en familia fuego fijo(s) (pasado o presente)			
Enstructura fija ocupada en fuego			
Reacción apropiada acerca del fuego encende			
Ausencias extensivas por padre			
Ausencias extensivas por madre			
Familia ha movido			
Huye de hogar			
Ha visto un consejero/terapeuta			
Otro miembro familiar ha visto un consejero/terapeuta			
Hechuras intentan a edad apropiado independencia de padres			
En problema en casa			
Padre o hermano con problema de la salud serio			
Matrimonio es infeliz			
Madre disciplina es en vigor			
Padre disciplina es en vigor			
Luchador con hermanos			
Conflictos en familia			
Fantasías raras			
Dibujos del pensamiento extraños			
Raro, ilógico, o lenguaje irracional			
Fuera de toque con realidad			
Calidad extraña acerca de niño			
Enojo de los expresos por herir mismo o algo que gusta			
Destruye propiedad propia			
Estaba/está en un culto			
Depresión severa o retiro			
Pobre or ningún ojo se pone en contacto			

Fineman, K , (1996) *Comprehensive Fire Risk Assessment.* Published in Poage, Doctor, Day, Rester, Velasquez, Moynihan, Flesher, Cooke, and Marshburn, (1997) Colorado Juvenile Firesetter Prevention Program: Training Seminar Vol 1 Denver, CO, Colorado Division of Firesafety   Parent FireRisk Questionnaire Page 4 of 5

RELEASE OF LIABILITY

I do hereby release, indemnify, and hold harmless the _____
Juvenile Firesetter Intervention Program, all its employees and volunteers against all claims, suits, or actions of any kind and nature whatsoever which are brought or which may be brought against the _____ Juvenile Firesetter Intervention Program for, or as a result of any injuries from, participation in this program.

_____   _____
            Parent/Guardian   Date/Time

_____   _____
                Juvenile   Witness

RELEASE OF CONFIDENTIAL INFORMATION

Juvenile's Name _____ D.O.B. _____

Release to/Exchange with:

    Name   _____

    Address _____

                 _____

    Phone   _____

Information Requested _____

_____

I consent to a release of information to and/or and exchange of information with the _____ Juvenile Firesetter Intervention Program. I understand that this consent may include disclosure of material that is protected by state law and/or federal regulations applicable to either mental health or drug/alcohol abuse or both.

This form does not authorize re-disclosure of medical information beyond the limits of this consent. Where information has been disclosed from records protected by Federal Law for drug/alcohol abuse records or by State Law for mental health records, federal requirements prohibit further disclosure without the specific written consent of the patient. A general authorization for release of medical or other information is not sufficient for these purposes. Civil and/or criminal penalties may attach for unauthorized disclosure of drug/alcohol abuse or mental health information.

A copy of this Release shall be as valid as the original.

_____   _____
      Parent/Guardian   Date/Time

_____   _____
         Juvenile   Witness

RISK ADVISEMENT

I have been informed that the FEMA/USFA Juvenile Firesetter Evaluation indicates that my child, _____ has a serious risk of continued involvement with fire setting activity.

I have also been informed by the _____ Juvenile Firesetter Intervention Program of the serious risk of injury and property damage that may continue to exist until the problem is resolved.

Appendix 4.1

Educational Package for
Juvenile Firesetter Intervention Programs

Program Guidelines	Educational Resources
Level 1	Preschool and Kindergarten (Ages 3 - 7)
Level 2	Grades 1- 3 (Ages 7 - 10)
Level 3	Grades 4 - 6 (Ages 11 - 13)
Level 4	Grades 7 - 12 (Ages 14 - 18)

Guidelines

A wide range of educational resource materials is available from several sources. These materials provide an excellent base for communities to draw upon when designing the educational component of their juvenile firesetter intervention programs. A listing of these resources by grade and age follows these guidelines.

Designing the Educational Intervention

Each juvenile firesetter program can develop an educational intervention designed to suit their needs. The exact format of the educational intervention depends on a variety of factors including the number of referrals, available resources, allotted time, severity of the firesetting, and the type of professional providing the intervention. The education component of the juvenile firesetter intervention program is usually provided by fire service professionals; but may also be provided in conjunction with mental health and social services, juvenile justice or other community agencies such as a hospital burn unit.

There is no prefect format or one ideal educational intervention that will work for all children. For example, some educational interventions provide weekly sessions for parents and children that are reinforced with homework assignments. In contrast, other programs offer one or two one-hour sessions in which more concise information is presented. The design of the educational intervention rests with each juvenile firesetter intervention program.

Who Participates?

At least one parent should be required to participate along with the child in all phases of the educational intervention. Ideally, both parents or caregivers should be present to demonstrate to their child a mutual concern and effort to extinguish the firesetting behavior. When all family members participate, the importance of the situation is clearly communicated. Often, parents learn as much as their children and can enforce safety awareness for the entire family. Siblings also should be encouraged to attend educational sessions, unless they are too young or they will detract from the educational program.

The Fire Service Response

The first respondent to a family crisis has a unique opportunity in many cases to make an unusually strong connection and to make a special impression. The fire safety educator, in addition to imparting sound information, can demonstrate through his or her interaction with family members good communication skills, problem solving techniques, consequences of behavior and respect for others.

The fire safety education session, handled with sensitivity and support, can lead the way to a positive course of action. The rapport established during those initial assessment and/or education sessions is essential to a successful intervention and referral.

When Is An Educational Intervention NOT Appropriate?

There will be times when education intervention should be delayed or only presented to parents. The following are examples of situations in which delay or elimination of education intervention for children is recommended.

1) ...the child and/or his family needs to be referred immediately to a mental health professional for further assessment and treatment.

This is a time when your efforts to develop a resource list and to establish a working relationship with mental health providers will serve you well. Knowing ahead of time the information needed by the clinician and their recognition of you as a person experienced in the field of juvenile firesetter assessment and intervention will make the process easier.

If possible, it is helpful to have an extensive list of referral sources. Geographic location, fee, sliding scales, and sex of therapist are often important to your client. It is helpful to provide parents with names, telephone numbers and addresses of the recommended therapists, and to make personal contact with the therapist informing them of the referral and your reasons for doing so. Check to make sure you have the necessary signed releases allowing information to flow in both directions, you to the therapist and she or he to you. Let the therapist know what you need to know for your records, i.e., did the child and his family follow through, goals of treatment, and when treatment is complete.

Communicate to the therapist that when the time is right, you are available to provide an education program for the child referred. Through collaboration, timing and content of the education session can be optimized.

If a child is in therapy, delay scheduling an education session until you have an opportunity to speak with the therapist and to discuss the advisability, timing, and content of your intervention.

2) …the child's ability to concentrate or comprehend is impaired to the degree that usefulness of the material presented will be minimal or misunderstood in a way that will be counterproductive.

In these circumstances, fire safety education for the parents is essential. Contact with the child's school or other care providers is an important safety issue. A support group for parents could be helpful.

3) …the child's fascination or excitement with fire will only be further enhanced by the fire education material. Referral to a mental health professional and fire safety education for parents is indicated.

4) …the very young child is better served by education and/or counseling for parents.

5) …the chronic juvenile offender. It is recommended that assessment and education intervention be discussed with the child's probation officer, juvenile diversion, DA, etc., to void reinforcing the behavior you are attempting to eliminate.

When Is An Educational Intervention Appropriate?

Educational intervention is almost always appropriate at some level, with the exceptions previously mentioned. The primary concern is that the information delivered is age-appropriate both in content and context for each child and family. The following is a recommended list of educational exercises by age categories. Also, following these educational interventions is a list of resources categorized by grade and age.

The Very Young Child (Ages 2 and 3)

There is growing concern about the number of very young children who are injured or killed as a result of fireplay and firesetting. This is a particularly troublesome age group due to the child's limited ability to:

    …understand the consequences of his behavior,
    …problem solve, and
    …appropriately respond once materials have ignited.

Firesetting and fireplay in this age group is usually a direct result of inadequate supervision or of the caretakers' failure to provide a safe environment. Education intervention for this age group is primarily focused on educating parents about THE BASICS, such as:

...fire tools and combustible materials should not be accessible to young children,
...children need constant adult supervision, and
...children are sensitive to environmental stress.

Very young children need constant supervision. Leaving children unattended while adult caretakers sleep, leave the premises or engage in activities without checking children's activities is dangerous. Teaching young children about danger is important; however, one must keep in mind that because their ability to learn is at a beginning level, lessons must be stated simply, must be repeated frequently, and immediately when the child attempts an unsafe behavior.

Parents are often lulled into complacency when children "are playing quietly in their room." Don't disturb the moment of tranquility! What they may not be aware of is that a frequent site of fireplay is under the bed or in a closet, secure hiding places for forbidden behavior with the potential for fatal consequences. Presenting statistics regarding the frequency of fire starts in particular locations in the home may establish the point with parents unaware of the dangers of fireplay in the home.

Even very young children respond to a stressful home environment. All children, but especially the very young, need structure, predictability, and a nurturing environment. Pre-school aged children have been known to act out their own stress with fireplay, especially when provided the opportunity due to poor supervision.

Parents determined to be experiencing psychosocial problems should be encouraged to seek counseling. It is wise to provide three names from which parents can choose. Most people seeking therapy or counseling select a therapist according to location, fee for service, confidence in the person making the referral and sometimes sex of the therapist. Providing more than one name gives the parents a piece of control in a circumstance in which they feel they have little.

In the main, children engaging in fireplay in this young age group are doing so out of experimentation, curiosity, and a drive to learn and imitate adult behavior. Education intervention, referral to parenting classes, or parent counseling is usually adequate. In some circumstances in which the level of chaos and family pathology is particularly evident, a referral to protective services and psychotherapy is indicated. It is important to emphasize to all parents that the young child is driven by normal curiosity and a passion for learning. This is clearly positive in a controlled, supervised setting involving safe learning materials. When the drive to learn is coupled with unsupervised use of fire materials, the results can be disastrous.

The Curious Child (Ages 3 and 4)

In the main, children engaging in fireplay in this young age group are doing so out of experimentation, curiosity, and a drive to learn and imitate adult behavior.

This type of child will benefit from working with the concepts of "tell an adult if you see matches or lighters," stop, drop, and roll, and familiarization with bunker gear. Children learn best from material that is presented to them from curricula that involves auditory, visual, and kinesthetic learning modalities in increments of twenty minutes or less.

Parents should be educated on the proper storage and use of ignition sources.

In cases where there is concern that the child's environment in the home is not safe, a referral to protective services is in order.

The Pre-School Child (Ages 4-5)

The concepts presented for the three and under child should be the primary focus of this age group as well. In addition, firesafety information should be included that is pertinent to the types of ignition sources that were used in the fireplay or firesetting, such as the use of candles, pilot lights, lamps, or flammable liquids. These children also are capable of participating in Exit Drills In The Home (EDITH), role play scenarios, and participating with their parents in evaluating their home for fire safety and survival.

Use of the same materials recommended for the three and under child is appropriate; however, the teaching techniques need to be age-appropriate. In addition, the Learn Not to Burn Resource Book provides good educational activities for concept reinforcement. Notify protective child services and/or provide a mental health referral as necessary.

The Kindergarten Child (Ages 5-7)

The following topics and exercises are age appropriate for this group of children. If indicated, referral can be made to child protective services or to mental health.

    Assisting with home fire safety inspections
    Burn Prevention/Care
    Fire Behavior
    Fire Survival Skills
    Fire Prevention Issues
    Peer Pressure
    Civil Liability

Elementary School Child (Ages 7-9)

The following topics and exercises are recommended for this age group. As needed, referral can be made to child protective services or to mental health.

    Assisting with home fire safety inspections
    Burn Prevention/Care
    Fire Behavior
    Fire Chemistry – Basic
    Flammable Liquid Properties
    Fire Survival Skills
    Fire Prevention Issues/Wildfire
    Peer Pressure
    Civil Liability

Children and Adolescents (Ages 10-18)

If the case has pending charges, please determine proper procedure per your individual juvenile justice agencies. The risk assessment and subsequent educational intervention should not be provided until the youth has completed appropriate juvenile justice procedures. If necessary, referral can be made to child protective services or to mental health.

Assisting with home fire safety inspections
Burn Prevention/Care
Fire Behavior
Fire Chemistry – Basic
Flammable Liquid Properties
Fire Survival Skills
Fire Prevention Issues/Wildfire
Peer Pressure
Civil Liability
Legal Liability – responsibility for damages
State Statutes
Arson, Reckless Endangerment, Firing of Woods and Prairies Statutes, Uniform Fire Code, and municipal ordinances
Security Clearance issues for future job applications

Child At Risk Of Continued Firesetting

If the child is at risk of continued firesetting, contact child protective services per local agreement, refer to mental health or report findings to the juvenile justice. Immediate referral to a mental health facility is necessary, and may require direct assistance from the risk assessment interviewer.

The parents need explicit fire survival information for their family to increase their level of awareness with regard to fire behavior. Educational intervention for the juvenile should be delayed until contact is made with a mental health professional. Inform the clinician of the various levels of education you can provide when the child has reached an appropriate point in their treatment. At that time, the content of the education session can be discussed.

It is recommended that the fire service offer a free home safety inspection to the family, including smoke detector check or installation. This should be completed when the children are not at home!

Educational Resources
Preschool and Kindergarten
(Ages 3 - 7)

Programs

Children's Television Workshop. Sesame Street. <u>Fire Safety Station</u>. New York: Children's Television Workshop, 1996. (English and Spanish) Audio tape included.

An activity book and audio tape designed to help educators reach preschoolers with six simple but essential fire safety lessons. This program uses the popular Sesame Street characters to deliver the fire safety messages.

> Contact: U.S. Fire Administration Publications
> 16825 South Seton Avenue
> Emmitsburg, MD 21727
> http://www.usfa.fema.gov

> Cost: Free

Kid Safe Program. Fire Safety Education Curriculum For Preschool Children. Oklahoma City Fire Department (1987).

An interactive, hands-on curriculum teaching nine fire safety lessons to preschoolers. Behavioral objectives, teaching outline, support activities and a video are some of the materials included in the program.

> Contact: Oklahoma City Fire Department
> Public Education
> 820 NW 5th Street
> Oklahoma City, OK 73106
> (405) 297-3314

First Step to Success. University of Oregon.

This program screens kindergartners for antisocial behavior. Those young children at risk receive a three-month program based on rewarding good behavior and showing parents, in their homes, how to teach their problem child to cooperate, make friends, and develop confidence.

Follow the Footsteps to Fire Safety Saint Paul Department of Fire and Safety Service (1998).

This is a prevention program for young children that uses the concept of "following the footsteps" to teach 10 fire safety lessons. Each footstep includes detailed lesson plans and sample worksheets. There also are materials for parents and teacher involvement.

> Contact: Paula Peterson
> (651) 223-6203

Learn Not to Burn® Preschool Program. English and Spanish (1997).

This program teaches fire safety awareness and skills to children ages 3 to 5 in group settings like day care centers or preschools. It includes lesson plans for eight observable behaviors, along with illustrations for coloring and worksheets, a cassette tape of songs, and information for parents and teachers.

      Contact: The National Fire Protection Association
                         1 Batterymarch Park
                         Quincy, MA 02269
                         (617) 770-3000
                         www.nfpa.org

Play Safe! Be Safe! Bic Corporation. Distributed by Fireproof Children.

This resource is focused on children ages 3 to 5 and includes a teacher's manual with four lesson plans, a videotape, with a series of interactive teaching tools, such as a colorforms set, story cards, and activity boards, and a card game.

      Contact: Fireproof Children
                         (716) 264-1754

The Safety Scholars. FIRE Solutions.®(1997)

This is a comprehensive curriculum for intervention education for children ages 3 to 7 that have played with or started fires. It includes interview forms, pre/post tests, lesson plans, worksheets, flannel board stencils, scripts and parent materials.

      Contact: FIRE Solutions
                         (508) 636-9149
                         www.firesolutions.com

Books and Materials

Baltimore County Fire Department. Carmen Sense.

These are fire safety learning lessons for preschoolers on a compact disk. This is an excellent interactive self-teaching tool. Contact: www.extrasense.com

Milton, Tony. Flashing Fire Engines. New York: Dutton Children's Books. (1999) $14.99.

This book follows three furred and feathered firefighters as they respond to calls for help. This material is meant to be read aloud. There is a wealth of basic information about firefighting equipment and procedures.

National Fire Protection Association (1995). Sparky's Fire Safety Coloring Book. Quincy, MA: National Fire Protection Association.

By coloring scenes, the young child learns seven fire safety rules.

Sis, Peter. <u>Fire Truck.</u> **New York: Greenwillow Books. (1999). $14.95.**

This story features a small boy who loves fire trucks so much that he awakens one morning to discover that he has become one. He travels around the neighborhood and rescues a cat, puts out a fire, and saves a teddy bear.

<u>**Brochures**</u>

For Parents

<u>Take 5 for Apartment Fire Safety</u>. **(1999). The St. Paul Fire Department.**

This is a checklist of potential fire hazards associated with apartment living. It also teaches seven basic fire safety behaviors for families and lists several important telephone numbers.

<u>Teaching Preschoolers To Be Fire Safe.</u> **(1994). Quincy, MA: National Fire Protection Association.**

This brief and colorful brochure describes why young children are at risk and need to be taught the basic rules of fire safety and survival.

<u>**Videos**</u>

<u>A Lighter Is Not a Toy.</u> **(1998). Quincy, MA: National Fire Protection Association.**

This eight-minute video includes several fictional vignettes that emphasize key match and fire safety messages. The video includes an instructional leader's guide and a reproducible handout highlighting key messages.

<u>Be Cool About Fire.</u> **Chicago, Illinois. Allstate Insurance Company.**

A short video emphasizing basic fire safety messages for preschoolers.

Educational Resources
Grades 1 - 3
(Ages 7 - 10)

Programs

Fireproof Children Education Kit. (1994). Pittsford, NY: National Fire Service Support Systems, Inc.

This program provides K-6 graders hands-on age-appropriate activities including songs, games, and experiments to teach fire safety and prevention.

> Contact: Fireproof Children
> 20 North Main Street
> Pittsford, New York 14534
> (716) 264-0840

Freddie Firefighter's Fire Safety and Burn Prevention Activity Packets. (1992). Plymouth, Minnesota: Genecom Group, Inc.

Freddie Firefighter has been sharing fire safety and burn prevention messages since 1975. This updated program, supported by the International Association of Fire Chiefs, Inc., stresses that parents and children must work together to learn the eight steps to fire safety. These steps are taught through the use of puzzles, games, and activities.

> Contact: Genecom Group, Inc.
> P.O. Box 47302
> Plymouth, Minnesota 55447
> (612) 559-7247

Learn Not To Burn Curriculum. K-8. (1997). National Fire Protection Association.

This widely used program teaches 25 key fire safety behaviors to K-8 classrooms through the use of goal-directed curriculum.

> Contact: National Fire Protection Association
> 1 Batterymarch Park
> Quincy, MA 02269
> (617) 770-3000

The Smoke Detective. (1990) Bloomington, Illinois: State Farm Insurance Companies.

This is a year-round program of fire safety designed for use in grades K - 6. It includes lesson plans, seasonal activities, worksheets, and a video.

> Contact: Smoke Detection
> State Farm Insurance Companies
> One State Farm Plaza
> Bloomington, Illinois 61710-0001

Books and Materials

Bridwell, Norman. <u>Clifford the Firehouse Dog.</u> Jefferson City, MO: Scholastic Press, 1995. ISBN #48419-2

Clifford, a big red dog familiar to most elementary school students, visits a fire station and helps firefighters put out fires. Fire safety messages highlighted in the book include stop, drop and roll, how to develop a home escape plan, the importance of checking smoke detector batteries, and the dangers of playing with matches.

Campbell, Chuck. <u>Sam's Big Decision.</u> Salem, Oregon: Oregon State Fire Marshal's Office, 1988.

This is a comic book that helps children talk about peer pressure by deciding what Sam should do when his friend wants to play with fire.

FIRE Solutions. <u>Fire Safety Flannel Board Stories.</u> Fleetport, MA: FIRE Solutions.

Five story scripts and over 52 illustrations teach children who are curious about fire what it is, how it works, and where it comes from. The emphasis is on teaching children to use fire in a positive and constructive way.

Johnston, Karen. <u>The Day Freddy's Bubble Burst.</u> Salem, Oregon: Oregon State Fire Marshal's Office, 1988.

This comic book helps children talk about their feelings before, during, and after a firesetting incident.

Muster Mouse Studios. <u>Muster Mouse Prevention Through Education.</u> Harrie, New York: Muster Mouse Studios, 1998.

This is a catalogue of fire safety and prevention books, activities, and materials for purchase based on a muster mouse theme.

National Fire Protection Association. <u>Sparky's ABCs of Fire Safety.</u> Quincy, MA: National Fire Protection Association.

Sparky the fire dog leads children on a magical journey through Alphabet Land. Each letter teaches a different life-safety lesson, spelling out the whole fire safety story from A to Z. Children discover how fires start, how to prevent them, and what to do if fire strikes. Sparky encourages children to join his fire prevention team and work together to win the fight against fire dangers.

<u>Safety Always Matters. Fire Safety Activity Book.</u> St. Rose, LA: Syndistar, Inc., 1992.

Fifteen fire safety lessons are included in this workbook which uses puzzles, games, drawing and other skills to teach children in grades K - 3.

Brochures

For Parents

<u>Big Fires Start Small.</u> (1996). Quincy, MA: National Fire Protection Association.

This colorful brochure focuses on the role of parents in helping to prevent child-set fires.

Juvenile Firesetters. What You Can Do. (1998) Emmitsburg, Maryland: National Arson Prevention Clearinghouse.

This brochure describes why children set fires and what parents can do to help.

Match and Lighter Fire Safety (1992) Quincy, MA: National Fire Protection Association.

Children, fire and basic fire safety rules for parents are the topics of this brochure.

Questions and Answers About Child-Resistant Lighters. Owensboro, KY Cricket B.V.

This brochure describes the child-resistant lighter, offers a diagram of how it works, and cautions parents to keep all lighters away from children.

Small Hands Big Fires. (1989). St. Rose, LA: Syndistar, Inc.

The profile of a child firesetter is presented with a checklist of behavioral symptoms for parents to answer yes or no, and suggestions for prevention and intervention.

United States Fire Administration (2001). Children and Fire...A Growing Concern

This brochure is directed toward educating parents and the community about the problem of juvenile firesetting. It also suggests several solutions. The brochure is free, and 200 copies can be ordered at a time at www.usafa.fema.gov

Videos

Donald's Fire Drill. Disney Educational Productions. (800) 295-5010.

Two students match wits and fire safety knowledge on the comical quiz show "Donald's Fire Drill" as they race to solve fire safety clues and questions based on Exit Drills In The Home (EDITH). Donald Duck demonstrates their answers.

Donald's Fire Survival Plan. Disney Educational Productions. (800) 295-5010.

Donald Duck and his nephews outline techniques to prevent or survive fire in the home. The program stresses the need for prevention, and presents stop, drop, and roll, and EDITH exit drills.

Educational Resources
Grades 4 - 6
(Ages 11 - 13)

Programs

Look Hot? Stay Cool! The Disaster Services Preparedness Bureau of the American Red Cross in collaboration with the St. Paul Department of Fire and Safety Service. (1998).

This juvenile firesetter prevention program consists of two sections, a youth unit designed for children ages 10-12, and an adult unit designed for parents and caregivers of children ages 10-12. There are key fire safety messages taught by the St. Paul Fire and Safety Service in collaboration with American Red Cross personnel, classroom teachers, and fire department personnel.

        Contact: Your Local Red Cross Chapter

Talking to Children About Fire. A Preventor's Guide. FIRE Solutions.

This is a manual and guidebook for fire educators who want to incorporate more fire science into their classroom prevention visits. Organized into three sections, for grades K-2, 3-4, and 5-6, the manual explains what children at each age level are capable of learning about fire and why it might hold such an appeal. There are lesson plans, teacher resource guides, and math and science activity sheets.

        Contact: FIRE Solutions
                        (507) 676-2334
                        www.firesolutions.com

Books and Materials

Accent Publishing. Junior Firefighter Activity Sheets. Portland, Oregon: Accent Publishing, 1992.

These are a series of age-appropriate activity sheets designed to teach such fire safety lessons as how to protect your family home, the science of fire, smoke detectors, and fire escape plans.

Cone, Patrick. Wildfire. Minneapolis: Carolrhoda Books, Inc. $7.95 paperback.

This book includes a brief history and well-illustrated description of fire in the wildland setting. It also refers to the role of fire in the ecosystem. Basic fire science information is presented with photographic illustrations.

Golden, Barbara. Coyote and the Fire Stick. New York: Gulliver Books (Harcourt Brace & Company). $15.

This book describes a Pacific Northwest legend about Coyote who steals fire from three evil spirits with the help of a mountain lion, deer, squirrel, and frog. The fire is swallowed by tree, but Coyote teaches people to recover fire by rubbing two sticks of wood together.

National Safe Kids Campaign. Safe Kids Are No Accident. A Fire Safety Booklet for Kids. (1991).

This is colorful workbook with games, fire facts, activities, and tips for children and their parents. Completion of the workbook elevates the child to Junior Fire Inspector!

Oregon State University Extension Service. Home Alone and Prepared. Prineville, Oregon: Oregon State University Extension.

This workbook and video, featuring Fireman Dave, is designed to teach children the necessary skills and knowledge to enable them to be safe and prepared when they are home alone. Six topics are covered, including determining the child's readiness for self-care, guidelines for making house rules, personal safety and plans when home alone, fire safety, first aid, and kitchen skills and food safety.

Rabbini, Ken. Fire. (The Elements, 3). New York: Henry Holt and Company, Inc. $16.95.

A scientific look at fire in all its forms. This book looks at fire's many roles as energy, heat, light, danger, and an element in rituals.

Scholastic Press. Taming Fire. Jefferson City, MO. ISBN#47637. $19.95.

This book follows fire through the earliest myths to Ben Franklin's experiments with lightning. Fire is explored throughout the world, from volcanoes to space shuttles to firefighting. This is an interactive book where you can turn a transparent page to see how a geyser works, how metal is forged, and how to make stained glass windows.

Simon, Seymour. Wildfire. New York: Marrow Junior Books. $15.

This book uses the Yellowstone fire of 1988 as well as fires in the Everglades to show that fire is both good and bad and is part of the life cycle.

Waldman, Larry. Who's Raising Whom. Phoenix, Arizona: UCS Press, 1994, $18.95 paperback.

This book is written for parents to help them understand why their children behave in certain ways and how to respond and manage their children's behavior.

Brochures

Fire Stoppers of Washington. A Family's Response to Firesetting. Seattle, Washington: Washington Insurance Council, 1997.

This is a parent education tool that can be used by fire and mental health professionals who work with juvenile firesetters. It is a 15-page booklet that presents information on factors that contribute to child fireplay, understanding what fire really is like, easy access to matches and lighters, parent and caregiver supervision, and the psychological factors associated with firesetting. The booklet also includes a plan of action that parents and caregivers can take to change fireplay and firesetting behavior.

      Contact: Washington Insurance Council
                         1904 3rd Ave., Suite 925
                         Seattle, Washington 98101-1123

International Shrine Headquarters. Burn Prevention Tips. (English and Spanish). Tampa, Florida: International Shrine Headquarters.

Eight critical burn situations are discussed in this 25-page booklet. Burn and fire prevention topics include kitchen safety, microwave burns, dangers of gasoline, home fires, match safety, first aid for burns, and camping, campfires, and grills.

Contact: Public Relations Department
            International Shrine Headquarters
            PO Box 31356
            Tampa, Florida 33631-3356

Phoenix Fire Department. Youth Firesetter Intervention Program. <u>A Parent's Guide.</u> Phoenix, Arizona: Phoenix Fire Department Youth Firesetter Prevention Program Team, 1998.

This booklet provides some of the warning signals parents need to be aware of concerning fireplay and firesetting. Parent responsibilities and tips are presented for three different age categories: children under seven, children age 8-12 years, and youth age 13-18 years. There is a suggested reading list for parents.

Contact: Youth Firesetter Intervention Program
            Phoenix Fire Department
            (602) 262-7757

St. Paul Department of Fire. <u>Inspect and Correct</u>. St. Paul, Minnesota: St. Paul Fire Department.

This booklet covers six important topics of fire prevention in the home. There is a fire safety checklist, along with information on smoke detectors, planning an escape, arson, fire prevention and public education, and paramedic-ambulance services. A list of important phone numbers also is included.

Contact: St. Paul Fire Department
            100 East Eleventh Street
            St. Paul, Minnesota 55101

The Children's Hospital Burn Center. <u>Fire, Kids, and Fire Setting.</u> Denver, Colorado: The Children's Hospital Association and the Colorado Juvenile Firesetter Prevention Program, 1997.

This booklet covers several topics on children and firesetting. It offers a brief explanation of the problem, a description of the warning signs, and encourages parents and caregivers to seek help. It also contains fire prevention and safety information for parents.

Contact: The Children's Hospital Burn Center
            1056 East 19th Avenue
            Denver, Colorado 80218
            (303) 861-6604

<u>Videos</u>

The Idea Bank. <u>In Their Own Words.</u> Santa Barbara, California: The Idea Bank. 1997. 15 minutes. $195.

This video portrays the story of three teenagers from different cities whose lives were changed by fire. It covers the emotional, financial, and legal price each teenager pays for setting arson fires. The video comes with lesson plans, an instructor's guide which includes the video script, and a resource guide.

Contact: The Idea Bank
            1139 Alameda Padre Sierra
            Santa Barbara, California 93103
            (800) 621-1136
            Fax (805) 965-2275
            e-mail info@theideabank.com

Educational Resources
Grades 7-12
(Ages 14-18)

Programs

Cooper, Traci. <u>P.A.L.S. Prevention Arson Loss in Schools.</u> Albany, Oregon.

This violence prevention program for middle schools presents six skill-based lessons. To evaluate its impact, students receive pre- and posttests. The program teaches decisionmaking skills and understanding the consequences of using fire inappropriately.

Elliot, Eric. <u>Skills Curriculum for Intervening With Firesetters.</u> Eugene, Oregon, 1997. 114 pages. $29.95.

This guide is divided into 14 lessons that help identify the underlying causes of juvenile firesetting. It is designed to be used by fire service personnel, mental health professionals, and parents.

> Contact: Eric Elliot
>   3150 Wayside Loop
>   Eugene, Oregon 97477
>   (541) 682-4742

FIRE Solutions. <u>The Science of Sizzle.</u> Fall River, MA: FIRE Solutions, Inc., 1996. $75.90.

This is a middle school fire science curriculum covering combustion, electricity, fire, natural gas, flammable liquids, fire in the environment, and the science of fighting fires.

> Contact: FIRE Solutions
>   PO Box 2888
>   Fall River, MA 02722
>   (508) 636-9149

Phoenix Associates. <u>Challenge for Life.</u>

The high school curriculum teaches 12 critical arson and fire prevention problems and solutions.

> Contact: Georgia Firefighters Burn Foundation
>   www.gfbf.org/challengerforlife

Books and Materials

St. Paul Department of Fire and Safety Services. <u>The Burn Problem.</u> Description. <u>Terms and Rehab.</u> St. Paul, Minnesota Department of Fire and Safety Services.

This is a 40-page document written for the older adolescent on burns and burn prevention. The topics include the physiology of a burn, and the classification of burn degrees, burn risk groups, burns in the kitchen, contact burns, hypothermia and frostbite, smoking materials, electrical injuries, and flammable and combustible materials.

Waldman, Larry. Coping with Your Adolescent. Norfolk, Virginia: Hampton Roads Publishing Company, Inc., 1994.

This book is written to help parents cope with their children during the teenage years. It also gives parents useful advice about how to help their teenagers through difficult situations.

Brochures

AEtna Life and Casualty. Fighting Back. A Community Guide to Arson Control.

This brochure describes the problem of arson, early warning signs, and what communities can do to prevent it.

Contact: AEtna Life and Casualty Corporate Communications, DA06
151 Farmington Avenue
Hartford, Connecticut 06156
(203) 237-3282

National Fire Protection Association. False Alarms and Arson. Quincy, MA: National Fire Protection Association, 1991.

False alarms, who turns them in, arson, who sets fires, and arson prevention are the topics covered in the brochure.

Contact: National Fire Protection Association
1 Batterymarch Park
Quincy, MA 02269-9101

Videos

Action Training Systems, Inc. Portrait of a Serial Arsonist. The Paul Keller Story, 1995. 50 minutes. $150.

This documentary includes interviews with Paul Keller, his defense attorney, his father who turned him in, and prosecutors. It portrays the obsession and pain behind the crime of arson.

Contact: Action Training Systems, Inc.
12000 NE 95th Street, #500
Vancouver, WA 98682
(800) 755-1440

Champaign Fire Department. Only a Minute to Learn, Only a Second to Burn. Champaign, Illinois: Champaign Fire Department. 12 minutes. $40.

This video uses a number of young burn survivors telling their own stories about the importance of information related to preventing and treating burns. It comes with a training outline that suggests one way to use the video with a middle school classroom.

Contact: Champaign Fire Department
307 S. Randolph
Champaign, Illinois 61820
(217) 351-4574

Insurance Federation of Minnesota. <u>Marked by Fire.</u> St. Paul, Minnesota: Insurance Federation of Minnesota. 1996. 20 minutes. $14.

This video tells the story of a young man serving a prison sentence for the crime of arson. It shows how a firesetting incident changed his life and the impact it had on his family and the victims of the fire.

> Contact: Insurance Federation of Minnesota
> 750 Northwest Center Tower
> 55 Fifth Street East
> St. Paul, Minnesota 55101
> (612) 292-1099

NOVA Video. <u>Hunt for the Serial Arsonist.</u> South Burlington, Vermont: NOVA Video, 1996. 60 minutes. $19.95.

This PBS documentary dramatically recounts the story of a Los Angeles fireman who is now serving a sentence for arson.

> Contact: NOVA Videos
> PO Box 2284
> South Burlington, Vermont 05407
> (800) 255-9424

<u>Brian's Story.</u> 1991. 15 minutes. $54.95.

Brian was a teenager when he was charged, prosecuted, and convicted of arson in Orinda, California. The fire he set destroyed six homes. The video was produced as part of Brian's sentence.

> Contact: Firefighter's Bookstore
> 18281 Gothard #105
> Huntington Beach, CA 92648
> (800) 727-3327

<u>Through the Eyes of a Child: Burn Recovery.</u> Denver, Colorado: The Children's Hospital Burn Center. 12 minutes. $60.

This video covers the physical, psychological, and social repercussions explained by children who have been burned and who are recovering from burns. The messages are delivered by the children themselves. Counselors and therapists also offer their observations.

> Contact: The Children's Hospital Burn Center
> 1056 East 19th Avenue
> Denver, Colorado 80218
> (303) 764-8295

Appendix 4.2

Follow-up Surveys

1. **Fire Stoppers (Two Formats)**
   King County, Washington

2. **Youth Fire Safety Program**
   Youth Firesetter Intervention Program
   Phoenix, Arizona

FIRE
Stoppers

Children's Fire
Prevention Program

Of King County

REFERRAL CLIENT 3 & 6 MONTH FOLLOWUP

FILING DATA
Case Number: _____   _____   _____

YEAR MONTH H   FDID# -   CONTACT#-

Child's Name: _____

Person Conducting Followup #1: _____   Date of Followup: _____
Person Conducting Followup #2: _____   Date of Followup: _____

Please circle the appropriate number.
You were referred to another agency. Did you go?

	Poor	Fair	Excellent

How would you rate the improvement (if any) in the child's
behavior since involvement with this program?

Emotionally?	1 2 3 4 5 NA
Continued use of fire?	Y ( ) N ( )
Overall?	1 2 3 4 5 NA

How consistent has your family been in keeping matches/lighters out of the child's environment?	1 2 3 4 5 NA

As a parent/guardian how satisfied were you with:

the fire safety education provided in counseling?	1 2 3 4 5 NA
the counselor's skills/rapport with the child and family?	1 2 3 4 5 NA
the overall counseling process?	1 2 3 4 5 NA

How would you rate the improvement (if any) in the child's
behavior since the last followup 3 months ago?

Emotionally?	1 2 3 4 5 NA
Continued use of fire?	1 2 3 4 5 NA
Overall?	2 3 4 5 NA

Does your family employ the fire safety education received in this program (i.e., test smoke detector, escape plan, etc.)?	1 2 3 4 5 NA

FIRE
Stoppers

Children's Fire
Prevention Program

Fax forms to Fire Stoppers at 206-296-7741

DEMOGRAPHIC CLIENT 3 MONTH FOLLOWUP

FILING DATA
Case Number: _____   _____   _____

    YEAR MONTH H   FDID# -   CONTACT#-

Child's Name: _____
Person Conducting Followup _____   **Date of Followup:** _____

Please circle the appropriate number.

| | Poor | Fair | Excellent |

How would you rate the improvement (if any) in the child's
behavior since involvement with this program?

Emotionally?   1 2 3 4 5 NA

Continued use of fire?   Y ( )   N ( )

Overall?   1 2 3 4 5 NA

How consistent has your family been in keeping
matches/lighters out of the child's environment?   1 2 3 4 5 NA

As a parent/guardian how satisfied were you with:

    the fire safety education provided for your child?   1 2 3 4 5 NA

    the interviewer's skills/rapport with the child and
    family?   1 2 3 4 5 NA

    the overall program?   1 2 3 4 5 NA

How would you rate the benefit of the fire safety education for your child?

Any additional comments or suggestions regarding this program: _____
_____
_____

YOUTH FIRE SAFETY PROGRAM
POSTCARD EVALUATION

PLEASE CHECKMARK 4 SELECTION

		YES	NO
1.	Has your child set a fire since he/she attended the fire safety class? * If answer is yes help is available, please call _____ .	o	o
2.	Did you practice your home escape plan?	o	o
3.	Have you checked your smoke detector since the fire safety class?	o	o
4.	Have you and your child talked about consequences of firesetting since the class?	o	o
5.	Do you keep lighters and matches out of the reach of children?	o	o

If your child participated in counseling offered by the Youth Fire Safety Program please answer the following question:

1.	Has there been any fire play or firesetting since the last counseling session?	o	o

Additional Comments: _____

_____

_____

(Enlarged sample)

Appendix 4.3

Restitution Agreement

RESTITUTION AGREEMENT

I, _____ agree to pay restitution to the victim of my fire in the amount of $ _____ . I will pay _____ per _____ until it is all paid.

The restitution will be paid to:

Name: _____

Address: _____

<table>
<tr><td>_____</td><td>_____</td></tr>
<tr><td>**Parent/Guardian**</td><td>**Date/Time**</td></tr>
<tr><td>_____</td><td>_____</td></tr>
<tr><td>**Juvenile**</td><td>**Witness**</td></tr>
</table>

Appendix 6.1

Public and Private Support
For
Juvenile Firesetter Programs

Public and Private Support for Juvenile Firesetter Programs

The following is a partial list of national, state, and local organizations that have a stake in supporting the efforts of juvenile firesetter programs. Many of these organizations can offer different types of help to juvenile firesetter programs, including training workshops, data collection, in-kind contributions, public awareness support, contracts and grants.

National Support

Public Sector

- **Alliance for Fire and Emergency Management**
- **American Red Cross**
- **Arson Alarm Foundation**
- **International Association of Arson Investigators**
- **International Association of Black Professional Fire Fighters**
- **International Association of Chiefs of Police**
- **International Association of Fire Chiefs**
- **International Association of Fire Fighters**
- **National Association of State Fire Marshals**
- **National Association of Town Watches**
- **National Crime Prevention Coalition**
- **National Education Association**
- **National Firesafety Educators**
- **National Fire Academy**
- **National SAFE KID's Coalition**
- **National Sheriff's Association**
- **National Volunteer Fire Council**
- **Shriners Burn Institutes**
- **United States Fire Administration**
- **United Way**

National Nonprofit Foundations
(Awarding grants to programs for at-risk youth)

- **Carnegie Corporation of New York**
  437 York,
  New York 10022
  (212) 371-3200
  www.carnegie.org

- John S. and James L. Knight
  One Biscayne Tower, Ste 3800
  2 S. Biscayne Blvd.
  Miami, FL 33131-1803
  (303) 539-0009
  www.knightfdn.org

- Lilly Endowment, Inc.
  2801 N. Meridan St.
  Indianapolis, Indiana 46208-0068
  (317) 924-5471

- Open Society Institute
  Center on Crime, Communities and Culture
  400 W. 59th St.
  New York, New York 10019
  (212) 548-0135
  www.soros.org/crime/

- Robert Sterling Clark Foundation, Inc.
  135 E. 64th St.
  New York, New York 10021
  (212) 288-8900
  www.rscf@aol.com

- The George Gund Foundation
  1845 Guildhall Bldg.
  45 Prospect Ave. W
  Cleveland, Ohio 44115
  (212) 241-3114
  www.gundfdn.org

- WK Kellogg Foundation
  1 Michigan Ave. E
  Battle Creek, Michigan 49107
  (616) 968-1611
  www.wkkf.org

Private Sector

- **Aetna Life and Casualty**
- **Allstate Insurance Company**
- **Children's Television Workshop**
- **Factory Mutual Insurance Company**
- **Insurance Committee for Arson Control**
- **Insurance Information Institute**
- **Laborers International Union**
- **National Fire Protection Association**
- **SOS Fires: Youth Intervention Programs**
- **State Farm Insurance Company**
- **The Idea Bank**
- **Walt Disney Enterprises**

State and Local Support

Public Sector--Community Organizations

- **Children's Hospitals and Burn Units**
- **Health and Social Services**
- **Members of the television, radio, and print media**
- **Parks and Recreation**
- **Red Cross, local chapters**
- **Service clubs, such as the Free Masons, Lions Clubs, and Elks Clubs**
- **Youth organizations, such as the Boys' and Girls' Clubs, Boy Scouts/Girl Scouts, YWCA, and YMCA**

Public Sector--Education

- **Head Start**
- **Parent Teacher Associations**
- **Parent Teacher Organizations**
- **Preschool and daycare providers**
- **School Boards**
- **Special Education**

Public Sector--State and Local Officials

- **Board of Supervisors or City Council**
- **Mayor's Office**
- **National Governor's Association**
- **National League of Cities**
- **Office of State House/Assembly Representatives**
- **Office of State Senators**
- **Regional Governor's Association**
- **State Fire Academies**
- **State Fire Marshal's Office**

Public Sector

- Automobile clubs and associations
- Chamber of Commerce
- Local branches of insurance companies
- Merchants Associations
- Private daycare, preschool, elementary, middle, and high schools

Appendix 6.2

Juvenile Firesetter Programs

Organizational Charts

1. **Fire Stoppers**
   King County, Washington

2. **Juvenile Firesetter Prevention Program**
   State of Colorado

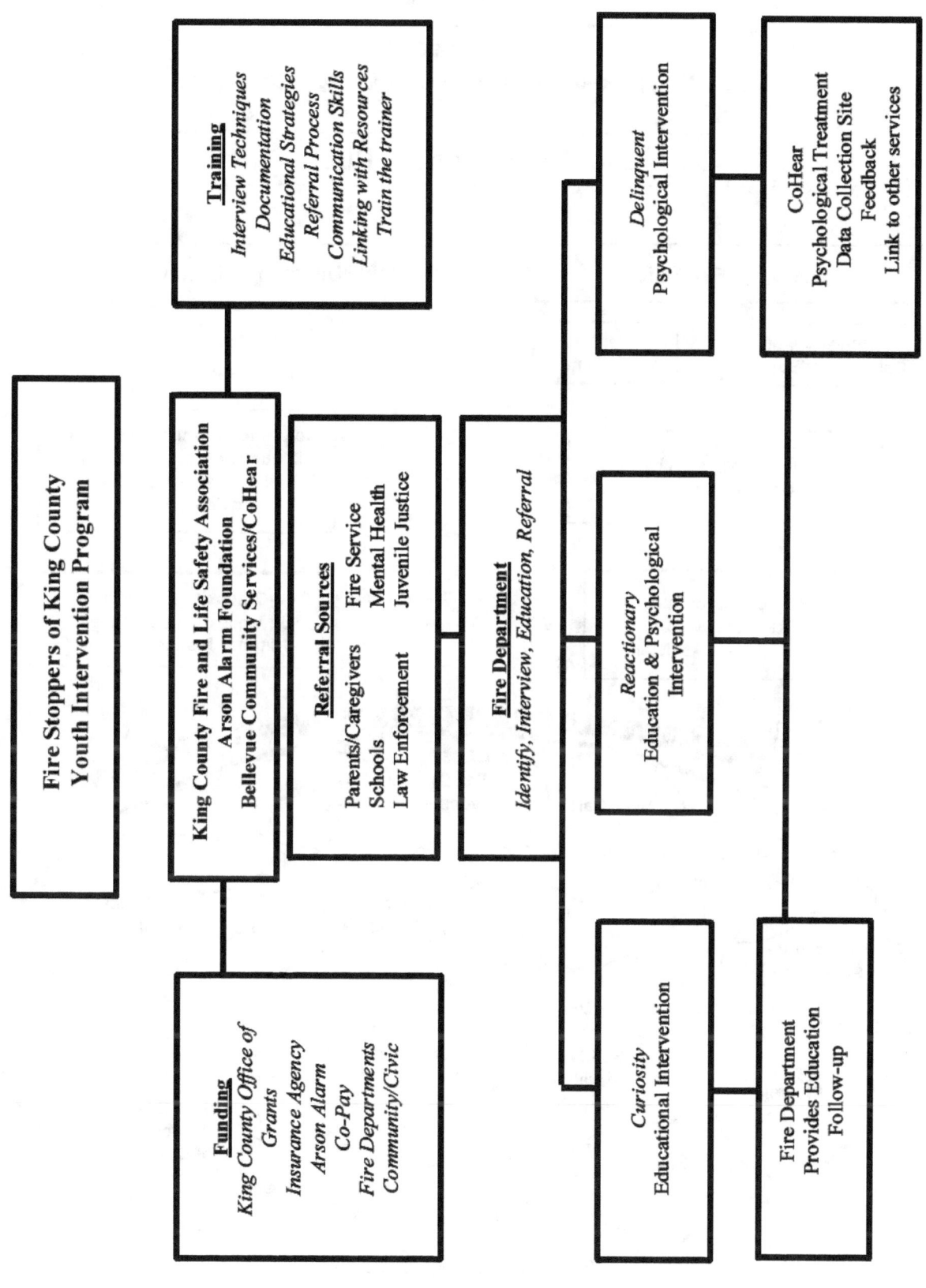

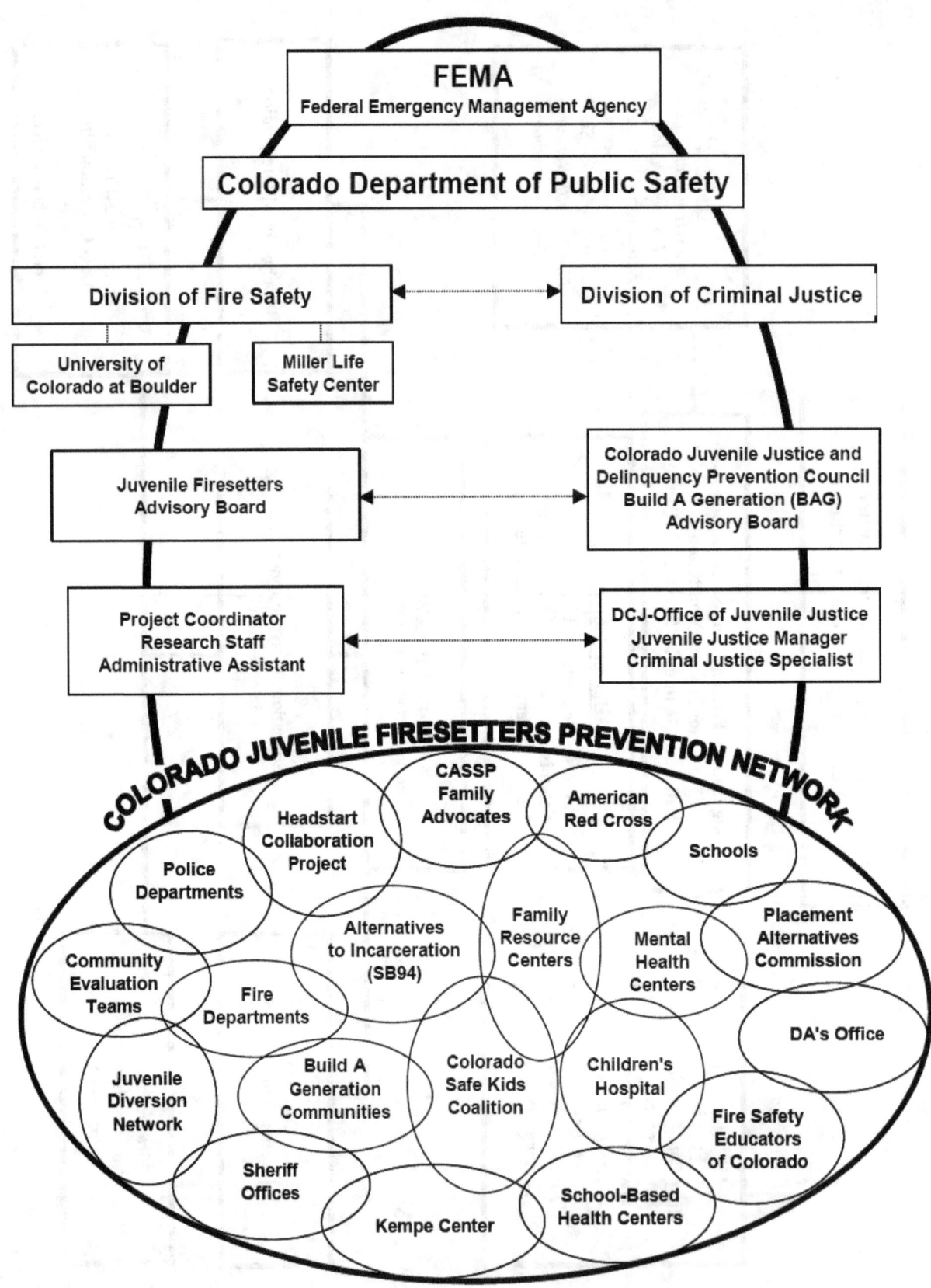

Appendix 6.3

FEMA/U.S. Fire Administration

CHILD FIRESETTING AND JUVENILE ARSON
Presentation/Training Package Series

1. Community Awareness Presentation

The materials presented in this package are designed to assist in creating a general, overall community awareness of the seriousness and magnitude of the child firesetting and juvenile arson problem as it exists nationally, and in your local area. Its purpose is to inform the community at large to better understand children's perception, use and misuse of fire. It is focused on motivating the community to organize and support a community-based prevention and intervention program to address this destructive fire problem with children and youth. Presented to a "town hall meeting" of community representatives, or to a local meeting of The Rotary International® for added support to your local program, the information offered will demonstrate that a community working together on this issue can make a difference!

This presentation can be delivered in a 2-4 hour time period (or less). The package includes: a presentation guide, a floppy disk with PowerPoint® presentation slides, sample PowerPoint® note-taking guide with resource information to duplicate for attendees, a sample brochure, logo sheet, and an 8 minute video titled **"Child Firesetting and Juvenile Arson--A Community Call To Action"**. Distribution will be through FEMA/U.S. Fire Administration--www.usfa.fema.gov

2. Professional Development Training Program

Materials include general information about child firesetting and juvenile arson, an overview of the development and implementation of a community-based program, and general principles applied to providing successful intervention. This package also addresses the more specific program management issues and support needed from related professional organizations and groups on a national, state, regional or local basis (State or Area Fire Chiefs Association, Arson Association State Chapters, Child Protective Services, Law Enforcement Agencies, etc.).

This program could be delivered at conference workshops, professional organization "schools" and inservices, and can range from 6-8 hours (or less) in duration. The package includes: an instructor guide, a floppy disk with PowerPoint® presentation slides, sample PowerPoint® note-taking guide with resource information to duplicate for attendees, a sample brochure, logo sheet, and an 8 minute video titled **"Child Firesetting and Juvenile Arson--A Community Call to Action"**. Distribution will be through FEMA/U.S. Fire Administration--www.usfa.fema.gov

CHILD FIRESETTING AND JUVENILE ARSON--PRESENTATION/TRAINING PACKAGE SERIES

3. Practitioner's Training Workshop

This training package includes instructional material, audio/visual aids, handout samples, and activities to conduct a two day, basic "how to" training session with professionals from all agencies who directly deal with firesetting children and youth. This includes, but is not limited to, the fire service, law enforcement, mental health, medical and education disciplines. In addition to the general information about the problem, and profile characteristics of firesetters, participants will actively review the interview risk screening process and forms, and identify useful interview techniques with children of all ages. The critical role of parents/caregivers in the success of the interview and intervention process is also discussed. The principles and elements of coordinating a community-based, multi-agency prevention and intervention task force, and local or regional program, including legal considerations, and evaluation will be addressed. A review of successful community programs, networking opportunities, and educational resources is included.

Workshop delivery time: two days (12 hours). The package includes: an instructor guide, a floppy disk with PowerPoint® presentation slides, black and white overhead transparency templates, sample PowerPoint® note-taking guide with resource information to duplicate for attendees, a sample brochure, logo sheet, an 8 minute video titled **Child Firesetting and Juvenile Arson--A Community Call To Action",** and a 10 minute video titled **"Interviewing Kids At Risk".** Distribution was accomplished in 1999-2000 to qualified instructors through ten train-the-trainer regional sessions throughout the United States.

Appendix 6.4

Web Pages and Internet Addresses

Web Pages and Internet Addresses

The topic of juvenile firesetting and arson can be explored on the Internet. Using a key word search of juvenile firesetting or arson, several items will surface. Below is a partial listing of the most frequently hit or requested sites on the Internet.

http://academic.uofs.edu/student/jn1/prop.htm

This contains an article on the relationship between parental smoking and the firesetting behavior of their offspring.

www.accessone.com/wa-ic/Jfsref.htm

This is a publication by the Washington Insurance Council titled, A Review of the Literature on Child Firesetters.

www.ci.phoenix.az.us/FIRE/firesetr.html

This site contains information on Phoenix, Arizona's Youth Firesetter Prevention Program.

www.fema.gov/napi/napiclh.htm

This is FEMA's National Arson Prevention Clearinghouse. It coordinates information about arson prevention resources. Several items are available including free arson prevention publications, and information about available grants, training, and public education programs.

www.usfa.fema.gov

This is the home page for the United States Fire Administration. Information on juvenile firesetting can be found on four lines: fire safety, arson prevention, publications, and the Kids Page.

www.fire.ottaw.on.ca/juvenile.html

This is the Juvenile Firesetter Intervention Program developed by Ottawa, ON, Canada.

www.firesolutions.com

This is FIRE solutions' web page that contains resource materials, such as juvenile firesetter intervention curricula, assessment guides, video tapes, brochures, and information on training for juvenile firesetter intervention programs.

www.juvenilejustice.com/

Juvenile Justice is an online, bi-monthly magazine serving juvenile justice professionals in all 50 states involved in youth service, human services, law enforcement, probation, parole, court administration, and staff training.

www.millersafetycenter.org

The Miller Safety Center has been providing juvenile firesetter intervention training and program development services to the U.S. Fire Administration and has been providing training in this field to state and local agencies since 1991. The Miller Safety Center and the Parker Fire Protection District are partners in prevention services. The Center provides consulting and training services in the area of new program development as well as training for mental health professionals, fire investigators, juvenile justice agencies, state agencies, NFPA 1035, and juvenile firesetter interventionists.

www.nfpa.org

This is the National Fire Protection Association's (NFPA) web page. It has all the information to access all areas of the NFPA's network, including code development, public education, conferences, and legislation. Links to all the national fire associations and organizations can be accessed. Also, the NFPA's education section and juvenile firesetter intervention committee can be accessed.

http://ojjdp.ncjrs.org

This is the site of the Office of Juvenile Justice and Delinquency Prevention. Among other things, they post the annual juvenile arrest records. In addition, they report on a wide range of topics related to juvenile crime.

www.osp.state.or.us/sfm/html/hot-issues.htm

Hot issues, a widely read quarterly newsletter of information and ideas on juvenile firesetting, is available online at this address. Hot Issues is published by the Oregon Office of the State Fire Marshal. Also at this address is information about the State of Oregon's Juvenile Firesetters Intervention Unit.

www.parkerfire.org

This organization hosts juvenile firesetter downloadable files for this manual and forms. Cheryl Poage's email address is: cpoage@parkerfire.org

www.pitt.edu/~kolko/fire.html

This site contains all the published articles on juvenile firesetting by David Kolko, Ph.D.

www.prisons.com/jo/arson.html

This is a brief article summarizing the work of the Office of Juvenile Justice and Delinquency Prevention by Eileen M. Garry titled, Juvenile Firesetting and Arson.

www.saic.com/firesafe/burn_10.html

This is the home page of the Burn Institute in San Diego.

www.sos.fires@fire.ci.portland.or.us

SOS FIRES is a non-profit advocacy agency for the development and maintenance of youth firesetting intervention programs. They maintain an interactive web site and provide training and consulting services for the development of new programs or maintenance of existing programs. They also provide media outreach and endeavor to provide a wide range of support services for any youth firesetting intervention efforts worldwide.

www.state.co.us/gov_dir/cdps/FireSafety/ProServ/cjfsp.htm

This is the home page for the Colorado Juvenile Fire Prevention Program.

www.theideabank.com

This is the Idea Bank's web page that contains a Resource Directory for Juvenile Firesetter Programs. It lists more than 200 resources in the United States and Canada. It is divided into four sections: people and programs, publications, video tapes, and web sites.

www.wa-ic.org/wic/juvfire.htm

This site contains information about the Washington Insurance Company's Juvenile Firesetter Program.